COMPUTATIONAL TECHNIQUES FOR ECONOMETRICS AND ECONOMIC ANALYSIS

Advances in Computational Economics

VOLUME 3

The titles published in this series are listed at the end of this volume.

Computational Techniques for Econometrics and Economic Analysis

edited by

D. A. Belsley

Boston College, Chestnut Hill, U.S.A.

Springer-Science+Business Media, B.V.

Library of Congress Cataloging-in-Publication Data

Computational techniques for econometrics and economic analysis / edited by David A. Belsley.
p. cm. -- (Advances in computational economics ; v. 3)
Includes index.

1. Econometric models--Data processing. 2. Economics, Mathematical--Data processing. I. Belsley, David A. II. Series.
HB141.C625 1993
330'.01'5195--dc20 93-17956

ISBN 978-90-481-4290-3 ISBN 978-94-015-8372-5 (eBook)
DOI 10.1007/978-94-015-8372-5

Table of Contents

Part Four: The Computer and Econometric Studies

DAVID A. BELSLEY, EDITOR

Preface

It is unlikely that any frontier of economics/econometrics is being pushed faster, further, and in more directions than that of computational techniques. The computer has become both a tool for doing and an environment in which to do economics and econometrics. Computational techniques can take over where theory bogs down, allowing at least approximate answers to questions that defy closed mathematical or analytical solutions. Computational techniques can make tasks possible that would otherwise be beyond human potential. And computational techniques can provide working environments that allow the investigator to marshal all these forces efficiently toward achieving desired goals.

This volume provides a collection of recent studies that exemplify all the elements mentioned above. And beyond the intrinsic interest each brings to its respective subject, they demonstrate by their depth and breadth the amazing power that the computer brings to the economic analyst. Here we see how modern economic researchers incorporate the computer in their efforts from the very inception of a problem straight through to its conclusion.

THE COMPUTER AND ECONOMETRIC METHODS

In "A Nonparametric Simulation Estimator for Nonlinear Structural Models," R. Bansal, A.R. Gallant, R. Hussey, and G. Tauchen combine numerical techniques, the generalized method-of-moments, and non-parametrics to produce an estimator for structural economic models that is defined by its ability to produce simulated data that best match the moments of a scoring function based on a non-parametric estimate of the conditional density of the actual data.

In "On the Accuracy and Efficiency of GMM Estimators: A Monte Carlo Study," A.J. Hughes Hallett and Yue Ma provide Monte Carlo evidence that helps to evaluate the relative small-sample characteristics of several of the more popular generalized method-of-moments estimators and surprise us by indicating that their own suggested method seems to work best.

In "A Bootstrap Estimator for Dynamic Optimization Models," A.J. Reed and C. Hallahan make use of a bootstrapping technique to provide estimates of stochastic, dynamic programming models that can be made to conform to boundary restrictions with relative ease.

In "Computation of Optimum Control Functions by Lagrange Multipliers," G. Chow explains and illustrates the gain in numerical accuracy that accompanies

D. A. Belsley (ed.), Computational Techniques for Econometrics and Economic Analysis, vii–ix.

his method of Lagrange multipliers for solving the standard optimal control problem over the more usual method of solving the Bellman equations.

THE COMPUTER AND ECONOMIC ANALYSIS

D. Kendrick in "Computational Approaches to Learning with Control Theory" discusses the means by which the more realistic assumptions that different economic agents have different knowledge and different ways of learning can be incorporated in economic modeling.

In "Computability, Complexity and Economics," A.L. Norman works to find a framework within which the mathematical theories of computability and complexity can be used to analyze and compare the relative merits of various of the optimizing procedures used in economics.

Berç Rustem, in "Robust Min-Max Decisions with Rival Models," provides an algorithm for solving a constrained min-max problem that can be used to produce a robust optimal policy when there are rival models to be accounted for.

COMPUTATIONAL TECHNIQUES FOR ECONOMETRICS

Continuing his tradition of seeing what the cutting edge has to offer economic and econometric analysis, W.L. Goffe, in "Wavelets in Macroeconomics: An Introduction," examines the usefulness of wavelets for characterizing macroeconomic time series.

C.R. Birchenhall, in "MatClass: A Matrix Class for C++," provides an introduction to object-oriented programming along with an actual C++ object class library in a context of interest to econometricians: a set of numerical classes that allows the user ready development of numerous econometric procedures.

In "Parallel Implementations of Primal and Dual Algorithms for Matrix Balancing," I. Chabini, O. Drissi-Kaïtouni, and M. Florian exploit the power of parallel processing (within the accessible and inexpensive "286" MS-DOS world) to bring the computational task of matrix balancing, both with primal and dual algorithms, more nearly into line.

THE COMPUTER AND ECONOMETRIC STUDIES

In "Variational Inequalities for the Computation of Financial Equilibria in the Presence of Taxes and Price Controls," A. Nagurney and J. Dong develop a computational procedure that decomposes large-scale problems into a network of specialized, individually-solvable subproblems on their way toward analyzing a financial model of competitive sectors beset with tax and pricing policy interventions.

In "Modeling Dynamic Resource Adjustment Using Iterative Least Squares," A. Somwaru, V.E. Bell, and U. Vasavada develop and illustrate a computational procedure for estimating structural dynamic models subject to restrictions such as the inequalities entailed on profits functions through convexity in prices.

Recognizing that corporate behavior can be greatly affected by qualitative as well as quantitative elements, A. Chakraborty and C.F. Baum, in "Intensity of Takeover Defenses: The Empirical Evidence," harness the power of the computer to allow them to study the qualitative issues surrounding the adoption and success of various anti-takeover devices.

David A. Belsley

PART ONE

The Computer and Econometric Methods

RAVI BANSAL, A. RONALD GALLANT, ROBERT HUSSEY AND GEORGE TAUCHEN

Computational Aspects of Nonparametric Simulation Estimation

ABSTRACT. This paper develops a nonparametric estimator for structural equilibrium models that combines numerical solution techniques for nonlinear rational expectations models with nonparametric statistical techniques for characterizing the dynamic properties of time series data. The estimator uses the the score function from a nonparametric estimate of the law of motion of the observed data to define a GMM criterion function. In effect, it forces the economic model to generate simulated data so as to match a nonparametric estimate of the conditional density of the observed data. It differs from other simulated method of moments estimators in using the nonparametric density estimate, thereby allowing the data to dictate what features of the data are important for the structural model to match. The components of the scoring function characterize important kinds of nonlinearity in the data, including properties such as nonnormality and stochastic volatility.

The nonparametric density estimate is obtained using the Gallant-Tauchen seminonparametric (SNP) model. The simulated data that solve the economic model are obtained using Marcet's method of parameterized expectations. The paper gives a detailed description of the method of parameterized expectations applied to an equilibrium monetary model. It shows that the choice of the specification of the Euler equations and the manner of testing convergence have large effects on the rate of convergence of the solution procedure. It also reviews several optimization algorithms for minimizing the GMM objective function. The Nelder-Mead simplex method is found to be far more successful than others for our estimation problem.

1. INTRODUCTION

A structural equilibrium model is a complete description of a model economy including the economic environment, the optimization problem facing each agent, the market clearing conditions, and an assumption of rational expectations. A structural equilibrium model is difficult to estimate, as doing so entails repeated solution of a fixed-point problem in many variables. One approach is to employ a linearization, typically linear-quadratic, in conjunction with Gaussian specification for the errors. A linear specification is attractive because a closed form solution can be obtained (Hansen and Sargent, 1980). However, recent advances in numerical techniques now make it possible to obtain good approximate solutions for nonlinear models. (See the 1990 symposium in the *Journal of Business and Economic Statistics* (*JBES*), summarized in Tauchen, 1990 and Taylor and Uhlig, 1990.) At the same time as these developments in structural modelling have occurred, purely statistical models, such as ARCH (Engle, 1982), GARCH (Bollerslev, 1986), and seminonparametric

D. A. Belsley (ed.), Computational Techniques for Econometrics and Economic Analysis, 3–22.

models (Gallant and Tauchen, 1989, 1992), have been used to discover and characterize important forms of nonlinear behavior in economic time series, especially in financial time series. Linear Gaussian models cannot explain such nonlinear behavior in actual data. Thus, nonlinear structural models must be examined to see the extent to which they can explain the nonlinear behavior found in actual economic data. This paper shows how statistical techniques can be combined with numerical solution techniques to estimate nonlinear structural equilibrium models.

The most common approach for estimation of nonlinear structural models is probably generalized method of moments (GMM) applied to Euler equations, as developed in Hansen and Singleton (1982). This technique has been widely employed in financial economics and macroeconomics, though it is a limited information method and has shortcomings. For example, the estimation can encounter problems when there are unobserved variables, as is the case for the model we consider in Section 2 where the decision interval is a week, but some of the data are observed monthly. Also it does not provide an estimate of the law of motion of the economic variables. Thus, if the model is rejected, little information is available regarding the properties of the observed data that the model has failed to capture.

In this paper we describe an alternative strategy for estimating nonlinear structural models that was first applied in Bansal, Gallant, Hussey, and Tauchen (1992). The approach is similar to the simulated method of moments estimators of Duffie and Singleton (1989) and Ingram and Lee (1991). However, unlike those estimators, which match preselected moments of the data, our estimator minimizes a GMM criterion based on the score function of a nonparametric estimator of the conditional density of the observed data. In effect, the estimator uses as a standard of comparison a nonparametric estimate of the law of motion of the observed data. By selecting the GMM criterion in this way, we allow the observed data to determine the dynamic properties the structural model must match.

The estimator works by combining the method of parameterized expectations for numerically solving a nonlinear structural equilibrium model (Marcet, 1991; den Haan and Marcet, 1990) with the seminonparametric (SNP) method for estimating the conditional density of actual data (Gallant and Tauchen, 1989, 1992). For a particular setting of the parameters of the structural model, the method of parameterized expectations generates simulated data that solve the model. The model parameters are then estimated by searching for the parameter values that minimize a GMM criterion function based on the scoring function of the SNP conditional density estimate. The nonparametric structural estimator thus has three components: (1) using SNP to estimate the conditional density of actual data, (2) using the method of parameterized expectations to obtain simulated data that satisfy the structural model, and (3) estimating the underlying structural parameters by using an optimization algorithm that finds those parameter values that minimize the GMM criterion function.

Below we discuss in detail how the estimator works in the context of a two-country equilibrium monetary model. The model is based on Lucas (1982), Svensson (1985), and Bansal (1990), and is developed in full detail in Bansal, Gallant, Hussey, and Tauchen (1992). It accommodates time non-separabilities in preferences (Dunn and Singleton, 1986) and money via a transactions cost technology (Feenstra, 1986). In effect, the model is a nonlinear filter that maps exogenous endowment and money supply processes into endogenous nominal processes, including exchange rates, interest rates, and forward rates. We show how this nonlinear dynamic model can be solved and simulated for estimation and evaluation.

In applying our estimator to this model, we find that there are several choices available to the researcher that greatly affect the estimator's success and rate of convergence. For example, the form in which one specifies the Euler equations on which the parameterized expectations algorithm operates can significantly affect the speed of convergence. This is an important finding, since our estimator uses this algorithm repeatedly at different model parameter values. Also, the means for testing convergence can have important consequences; we find it best to test for convergence of the projection used in parameterized expectations instead of testing for convergence of the coefficients representing the projection. Finally, we find that the complexity of our estimation procedure causes some optimization algorithms to have greater success in minimizing the GMM objective function. Among the optimization techniques we tried are gradient search methods, simulated annealing, and simplex methods. In Section 3.1 below we discuss how these methods work and their strengths and weaknesses for our type of optimization problem.

The rest of the paper is organized as follows: Section 2 specifies the illustrative monetary model and describes the simulation estimator. Section 3 discusses practical aspects of implementing the estimator, including solving the model with parameterized expectations and optimizing the GMM objective function to estimate the model parameters. Concluding remarks comprise the final section.

2. THE NONPARAMETRIC STRUCTURAL ESTIMATOR

2.1. *The Structural Model*

We apply our nonparametric structural estimator to the equilibrium monetary model of Bansal, Gallant, Hussey, and Tauchen (1992). In that model, a representative world consumer has preferences defined over services from two consumption goods. The utility function is assumed to have the form

$$E_0 \sum_{t=0}^{\infty} \beta^t \left[\left(c_{1t}^{*\,\delta} c_{2t}^{*\,1-\delta} \right)^{1-\gamma} - 1 \right] / (1-\gamma),$$

where $0 < \beta < 1$, $0 < \delta < 1$, $\gamma > 0$, and where c_{1t}^* and c_{2t}^* are the consumption services from goods produced in countries 1 and 2, respectively. Preferences are of the constant relative risk aversion (CRRA) type in terms of the composite consumption goods. The parameter γ is the coefficient of relative risk aversion, δ determines the allocation of expenditure between the two services, and β is the subjective discount factor. If $\gamma = 1$, then preferences collapse to log-utility

$$E_0 \sum_{t=0}^{\infty} \beta^t \left(\delta \ln c_{1t}^* + (1-\delta) \ln c_{2t}^* \right).$$

The transformation of goods to services is a linear technology

$$c_{1t}^{*} = c_{1t} + \kappa_{11}c_{1,t-1} + .. + \kappa_{1L_c}c_{1,t-L_c},$$

$$c_{2t}^{*} = c_{2t} + \kappa_{21}c_{2,t-1} + .. + \kappa_{2L_c}c_{2,t-L_c},$$

where c_{1t} and c_{2t} are the acquisitions of goods, the κ_{ij} determine the extent to which past acquisitions of goods provide services (and hence utility) in the current period, and L_c is the lag length. If $L_c = 0$, then the utility function collapses to the standard time separable case where $c_{1t}^{*} = c_{1t}$ and $c_{2t}^{*} = c_{2t}$. If the nonseparability parameters κ_{ij} are positive, then past acquisitions of goods provide services today. If they are negative, then there is habit persistence. Other patterns are possible as well. Recent acquisitions of goods can provide services today, while acquisitions further in the past contribute to habit persistence.

We introduce money into the model via a transaction-costs technology. The underlying justification for transactions costs is that the acquisition of goods is costly both in terms of resources and time. Money, by its presence, economizes on these costs and hence is valued in equilibrium. Transaction costs, $\psi(c, m)$, in our model are an increasing function of the amount of goods consumed c and a decreasing function of the magnitude of real balances m held by the consumer in the trading period. The functional form we use for the transaction-costs technology is

$$\psi(c, m) = \psi_0 c^{\alpha} m^{1-\alpha},$$

where $\psi_0 > 0$ and $\alpha > 1$.

The consumer's problem is to maximize expected utility $E_0 \Sigma_{t=0}^{\infty} \beta^t U(c_{1t}^{*}, c_{2t}^{*})$ by choosing $c_{1t}, c_{2t}, M_{1,t+1}, M_{2,t+1}, b_{1,t+1}^{k}$, and $b_{2,t+1}^{k}$, $k = 1, \ldots, N_a$, at time t subject to a sequence of budget constraints

$$\begin{aligned} P_{1t}[c_{1t} + \psi(c_{1t}, m_{1t})] &+ e_t P_{2t}[c_{2t} + \psi(c_{2t}, m_{2t})] \\ &+ \sum_{k=1}^{N_a}(1/R_{1t}^{k})b_{1,t+1}^{k} + \sum_{k=1}^{N_a}(f_t^{k}/R_{1t}^{k})b_{2,t+1}^{k} + M_{1,t+1} + e_t M_{2,t+1} \\ \leq &\sum_{k=1}^{N_a}(1/R_{1t}^{k-1})b_{1t}^{k} + \sum_{k=1}^{N_a}(f_t^{k-1}/R_{1t}^{k-1})b_{2t}^{k} + M_{1t} + e_t M_{2t} \\ &+ P_{1t}w_{1t} + e_t P_{2t} w_{2t} + q_{1t} + e_t q_{2t}. \end{aligned}$$

Here, P_{1t} and P_{2t} are current prices of consumption goods c_{1t} and c_{2t} in the units of the respective country's currency. $M_{1,t+1}$ and $M_{2,t+1}$ are the stocks of currency in the two countries carried forward from period t to $t+1$. Real money balances, $m_{1t} = M_{1t}/P_{1t}$ and $m_{2t} = M_{2t}/P_{1t}$, are defined in terms of beginning of period money holdings. The $b_{1,t+1}^{k}$ and $b_{2,t+1}^{k}$ are the agent's holdings of risk-free claims to the currencies of countries 1 and 2 in period $t+k$. Claims on country 1's currency are made by trading pure discount bonds with gross k-period interest rates R_{1t}^{k}. Claims on country 2's currency are made by trading forward contracts in the currency market, where e_t is the spot exchange rate and f_t^k is the k-period forward exchange rate, with both rates defined in units of country 1's currency per unit of country 2's currency. w_{1t} and w_{2t} are the stochastic endowments of goods within the two countries. Lump sum transfers of q_{1t} and q_{2t} units of currency are made by the government at time t. These transfers are known to the agent at the beginning of period t but can be used for carrying out transactions only in period $t+1$.

The stationary decision problem facing the agent delivers the following Euler equations for the asset holdings $M_{1,t+1}$ and $M_{2,t+1}$:

$$E_t\Big[MU_{c_{it}} - \beta MU_{c_{i,t+1}}\Big(\frac{P_{it}}{P_{i,t+1}}\Big)\Big(\frac{1+\psi_{c_{it}}}{1+\psi_{c_{i,t+1}}}\Big)\Big(1-\psi_{m_{i,t+1}}\Big)\Big] = 0, \quad i = 1,2,$$

and for $b_{1,t+1}^k$ and $b_{2,t+1}^k$:

$$E_t\Big[MU_{c_{1t}} - \beta^k MU_{c_{1,t+k}}\Big(\frac{P_{1t}}{P_{1,t+k}}\Big)\Big(\frac{1+\psi_{c_{1t}}}{1+\psi_{c_{1,t+k}}}\Big)R_{1t}^k\Big] = 0,$$

$$E_t\Big[MU_{c_{2t}} - \beta^k MU_{c_{2,t+k}}\Big(\frac{P_{2t}}{P_{2,t+k}}\Big)\Big(\frac{1+\psi_{c_{2t}}}{1+\psi_{c_{2,t+k}}}\Big)\Big(e_t R_{1t}^k/f_t^k\Big)\Big] = 0,$$

where $MU_{c_{it}}$ is the marginal utility of c_{it}, and $\psi_{c_{it}}$ and $\psi_{m_{it}}$ are the derivatives of transaction costs, $\psi(c_{it}, m_{it})$, with respect to the first and second arguments, respectively. Transactions costs modify the returns to the two monies, M_{1t} and M_{2t}. We would expect $P_{1t}/P_{1,t+1}$ to be the return at time $t+1$ for carrying forward an extra unit of country one's currency today. However, because of transaction costs, every extra unit of currency carried forward also lowers transaction costs in the next period by a real amount, $-\psi_{m_{i,t+1}}$, so the total return is given by $[(1-\psi_{m_{i,t+1}})P_{it}/P_{i,t+1}]$. The model also delivers an intratemporal restriction on the choice of goods c_{1t} and c_{2t}

$$e_t = E_t\Big[\Big(\frac{MU_{c_{2t}}}{MU_{c_{1t}}}\Big)\Big(\frac{P_{1t}}{P_{2t}}\Big)\Big(\frac{1+\psi_{c_{1t}}}{1+\psi_{c_{2t}}}\Big)\Big].$$

In maximizing utility, the consumer faces an exogenous stochastic process that governs the evolution of money growth and endowment growth in the two countries. We define the operator d to produce the ratio of the value of a variable in one period to its value in the previous period, as, for example, $dM_{1t} = M_{1t}/M_{1,t-1}$. Using this operator, we specify a driving process for the exogenous state vector $S_t = (dM_{1t}, dM_{2t}, dw_{1t}, dw_{2t})$ of the form

$$\log S_t = a_0 + A \log S_{t-1} + u_t,$$

where u_t is *iid* $N(0, \Omega)$, a_0 is a 4-vector, and A and Ω are 4×4 matrices. More complex stochastic processes for the exogenous state variables could easily be accommodated by our numerical solution method.

The final elements needed to complete the description of the model are the market clearing conditions

$$c_{it} + \psi(c_{it}, m_{it}) = w_{it},$$

$$M_{i,t+1} = q_{it} + M_{it} \,, \qquad i = 1, 2.$$

The parameter vector of the structural economic model is

$$\lambda = (\beta, \gamma, \delta, \psi_0, \alpha, \kappa_{11}, \ldots, \kappa_{1L_c}, \kappa_{21}, \ldots, \kappa_{2L_c}, a_0', \; vec(A)', \; vech(\Omega^{1/2})')'.$$

For each value of λ the model defines a nonlinear mapping from the strictly exogenous process $\{S_t\}$ to an output process $\{U_t\}$. The output process is

$$U_t = (dM_{1t}, dM_{2t}, dw_{1t}, dw_{2t}, dc_{1t}, dc_{2t}, dP_{1t}, dP_{2t}, R_{1t}^4, f_t^4/e_t, de_t)',$$

which is an 11-vector containing the elements of S_t along with the gross consumption growth rates, the gross inflation rates, the four-period interest rate in country 1, the ratio of the four-period forward exchange rate to the spot rate, and the gross growth rate of the spot exchange rate. It proves convenient also to include the elements of S_t in the output process, mapping them directly with an identity map. The particular set of variables comprising the remaining elements of U_t are those endogenous variables that turn out to be of interest for various aspects of the analysis of the model and the empirical work.

The mapping from $(\{S_t\}, \lambda)$ to the endogenous elements of U_t is defined by the solution to the nonlinear rational expectations model. In practice, we use Marcet's method of parameterized expectations (Marcet, 1991; den Haan and Marcet, 1990) to approximate the map. Given a value of λ, the method "solves" the model in the sense of determining simulated realizations of the variables that satisfy the Euler equations. In what follows, $\{U_t^\lambda\}$ denotes a realization of the output process given λ and a realization of $\{S_t\}$. A complete description of how we apply the method of parameterized expectations to this problem is given in Section 3.1 below.

2.2. *The Estimation Method*

The nonlinearity of the economic model prevents estimation by traditional methods since it is computationally intractable to compute the likelihood of a sample as a function of the model's parameters. However, simulation methods can be used to compute predicted probabilities and expectations under the model. Thus we propose a new simulation estimator that estimates the model by searching for the value of the parameter λ for which the dynamic properties of data simulated from the model match, as closely as possible, the properites of actual data.

Not all elements of U_t generated by the model are actaully observed weekly, so our empirical strategy is to use latent-variable methods with our simulation estimator. High quality observations on financial market prices, i.e., payoff data, are widely available on a weekly basis, and so we concentrate on these series in the estimation. We utilize weekly observations on three raw series: SPOT_t, the spot exchange rate (in \$ per DM); FORWARD_t^4, the 30-day forward rate (in \$ per DM); and TBILL_t^4, the one month treasury bill interest rate, computed from the term structure, and quoted on a bank discount basis. From the raw series we form a 3-element process $y_t = (y_{1t}, y_{2t}, y_{3t})'$ with

$$y_{1t} = 100 * \log(\text{SPOT}_t/\text{SPOT}_{t-1}),$$

$$y_{2t} = 100 * \log(\text{FORWARD}_t^4/\text{SPOT}_t),$$
$$y_{3t} = \text{TBILL}_t^4.$$

Exploratory empirical work indicates that $\{y_t\}$ is reasonably taken as a strictly stationary process, while the levels of the exchange rate series are nonstationary.

The correspondence between the elements of y_t and those of the output vector U_t are as follows: Country 1 is the U.S. and country 2 is Germany. Given a simulated realization $\{U_t^\lambda\}$ from the model, the corresponding $\{y_t^\lambda\}$ is computed as

$$y_{1t}^\lambda = 100 * \log(de_t^\lambda),$$
$$y_{2t}^\lambda = 100 * \log(f_t^4/e_t),$$
$$y_{3t}^\lambda = 100 * (360/30)[1 - (1/R_{1t}^{4\lambda})].$$

The expression for y_{3t}^λ converts $1/R_{1t}^{4\lambda}$, which is the price at time t of \$1 in period $t + 4$, to an annualized interest rate using the bank discount formula customarily applied to treasury bill prices (Stigum, 1990, p. 66).

The observed process is $\{y_t\}$ and the simulated process is $\{y_t^\lambda\}$ as defined above. The $\{y_t\}$ process is computed directly from the raw data while $\{y_t^\lambda\}$ is computed using the structural model of Section 2.1. We assume the model to be "true" in the sense that there is a particular value, λ_0, of the structural parameter vector and a realization, $\{S_{0t}\}$, of the exogenous vector such that the observed $\{y_t\}$ is obtained from $(\{S_{0t}\}, \lambda_0)$ in exactly the same manner that the model generates $\{y_t^\lambda\}$ from $(\{S_t\}, \lambda)$.

In broad terms, the estimation problem of this paper is analogous to the situation described, among others, by Duffie and Singleton (1989) and Ingram and Lee (1991). Common practice in such situations is to use a simulated method of moments estimator of λ_0 based on certain *a priori* selected moments of the data. We likewise propose such an estimator, but we take a different approach in determining what moments to match and in assigning relative weights in matching those moments.

The estimation strategy of this paper starts from the point of view that the structural model should be forced to confront all empirically relevant aspects of the observed process. The observed process $\{y_t\}$ is strictly stationary and possibly nonlinear, so its dynamics are completely described by the one-step ahead conditional density $f(y_t|\{y_{t-j}\}_{j=1}^\infty)$. Let $\hat{f}(\cdot|\cdot)$ denote a consistent nonparametric estimate computed from a realization $\{y_t\}_{t=t_0}^n$. The estimator $\hat{f}(\cdot|\cdot)$ defines what is empirically relevant about the process and thereby provides a comprehensive standard of reference upon which to match the economic model to the data.

The keystone to our structural estimator is the scoring function of the SNP estimator of Gallant and Tauchen (1989, 1992), which provides a consistent nonparametric estimator of the conditional density under mild regularity conditions. This use of the nonparametric fit to define the criterion of estimation motivates our choice of the term "nonparametric structural estimator". The Gallant-Tauchen estimator is a truncation estimator based on a series expansion that defines an hierarchy of increasingly complex models. The estimator $\hat{f}(\cdot|\cdot) = f_K(\cdot|\cdot, \hat{\theta}_{K_n})$ is characterized by an auxiliary parameter vector $\hat{\theta}_{K_n}$ that contains the coefficients of the expansion; the subscript K denotes the Kth model in the hierarchy. The length of $\hat{\theta}_{K_n}$ depends on the model. In practice, K is determined by a model selection criterion that slowly expands the model with sample size n and thereby ensures consistency. For the Kth model in the hierarchy, the corresponding $\hat{\theta}_{K_n}$ solves the first-order condition

$$\frac{\partial}{\partial \theta_{K_n}} \mathcal{L}_{K_n}(\{y_t\}_{t=t_0}^{n}, \hat{\theta}_{K_n}) = 0,$$

where $\mathcal{L}_{K_n}(\cdot)$ is the sample log likelihood of the corresponding model.

The nonparametric structural estimator is defined by mimicking this condition. Specifically, subject to identifiability conditions, a consistent estimator is available by choosing $\hat{\lambda}$ to make the same condition hold (as closely as possible) in the simulation

$$\frac{\partial}{\partial \theta_{K_n}} \mathcal{L}_{K_n}(\{y_\tau^{\lambda}\}_{\tau=\tau_0}^{\mathcal{T}}, \hat{\theta}_{K_n}) \approx 0.$$

The left-hand side is the gradient of the log likelihood function evaluated at a simulated realization $\{y_\tau^{\lambda}\}_{\tau=\tau_0}^{\mathcal{T}}$ and at the $\hat{\theta}_{K_n}$ determined by fitting the Kth SNP model to the actual data $\{y_t\}_{t=t_0}^{n}$. If the length of λ, ℓ_λ, is less than the length of θ_K, ℓ_K, then the model is overidentified (under the order condition) and a GMM criterion is used to minimize the length of the left-hand side with respect to a suitable weighting matrix.

Interestingly, this approach defines a consistent and asymptotically normal estimator irrespective of the particular SNP model used, so long as $\ell_K \geq \ell_\lambda$ and an identification condition is met. In practice, we implement the estimator using the particular SNP model that emerges from the specification search in the nonparametric estimation of $f(\cdot|\cdot)$. The choice of K is thus data-determined. This selection rule forces the scoring function to be appropriate for the particular sample at hand. The scoring function of the fitted SNP model contains just those indicators important to fit the data and no more. Also, because the fitted SNP model has the interpretation of a nonparametric maximum-likelihood estimator, the information equality from maximum likelihood theory provides a convenient simplification that greatly facilitates estimation of the weighting matrix for the GMM estimation.

3. IMPLEMENTING THE ESTIMATOR

In this section we discuss the practical aspects of implementing the nonparametric structural estimator described above. The implementation entails an initial SNP estimation of the conditional density of observed payoff data. The score function from this density estimate defines what properties our nonparametric structural estimator must mimic. Because estimating SNP models has been described extensively in Gallant and Tauchen (1989, 1992), we do not review that procedure here.

Following the SNP estimation, there are three distinct components to the procedure. The first involves using the method of parameterized expectations to solve the structural model for a particular value of the parameter vector λ. The second entails combining the initial SNP estimation with the parameterized expectations procedure to form the GMM objective function for the nonparametric structural estimator. The third is optimization of the objective function. Each of these components is described in detail below.

3.1. *Solving the Model Using Parameterized Expectations*

We use the method of parameterized expectations (Marcet, 1991; den Haan and Marcet, 1990) to obtain simulated data that satisfy the Euler equations of the structural economic model. In essence, this method approximates conditional expectations of certain terms with the projections of those terms on a polynomial in the state variables. The method uses Euler equations to iterate between postulated values of time series and projections based on those postulated values until those values and projections each converge. This procedure will be explained more fully below.

We find that the specification of the Euler equations greatly affects the speed with which the parameterized expectations algorithm converges. From Section 2, the first two Euler equations are

$$E_t\Big[MU_{c_{it}} - \beta MU_{c_{i,t+1}}\Big(\frac{P_{it}}{P_{i,t+1}}\Big)\Big(\frac{1+\psi_{c_{it}}}{1+\psi_{c_{i,t+1}}}\Big)\Big(1-\psi_{m_{i,t+1}}\Big)\Big] = 0, \quad i = 1,2.$$

Using the definition of the velocity of money, $V_{it} = c_{it}P_{it}/M_{it}$, $i = 1, 2$, one form in which these equations can be rewritten is

$$V_{it} = \frac{E_t\Big(MU_{c_{it}}\Big)}{E_t\Big[\beta MU_{c_{i,t+1}}\Big(\frac{dc_{i,t+1}}{dM_{i,t+1}V_{i,t+1}}\Big)\Big(\frac{1+\psi_{c_{it}}}{1+\psi_{c_{i,t+1}}}\Big)\Big(1-\psi_{m_{i,t+1}}\Big)\Big]}, \quad i = 1,2.$$

Because of the time nonseparabilities in our model, it is also possible to rearrange these Euler equations into an alternative form that expresses velocity as a single conditional expectation rather than the ratio of two conditional expectations. (We omit the derivation here.) It would seem at first that expressing the Euler equations as a single conditional expectation would be advantageous since the solution algorithm would have to estimate only one conditional expectation per Euler equation rather than two. However, we have found that convergence of the algorithm with this specification is much slower. This occurs because the single conditional expectation contains a difference of two terms that remains stable across iterations, while the time series from which it is constructed moves around substantially. The conditional expectation of this difference is less informative for updating guesses at the solution time series than are the two conditional expectations specified in the ratio above.

The next step in setting up the Euler equations entails various mathematical manipulations that allow them to be expressed in terms of conditional expectations of functions of velocity, consumption growth, and money growth:

$$V_{it} = \frac{E_t\Big[f_{i1}\Big(dc_{1,t-L_c+1}, dc_{2,t-L_c+1}, \ldots, dc_{1,t+L_c}, dc_{2,t+L_c}; \lambda\Big)\Big]}{E_t\Big[f_{i2}\Big(dc_{1,t-L_c+2}, dc_{2,t-L_c+2}, \ldots, dc_{1,t+L_c+1}, dc_{2,t+L_c+1}, V_{it}, V_{i,t+1}, dM_{i,t+1}; \lambda\Big)\Big]}$$

$$= \frac{E_t(F_{i1,t})}{E_t(F_{i2,t})} \qquad i = 1, 2,$$

where the $f_{ij}(\cdot)$ are particular functional forms too complex to be written out here. The market clearing conditions of the model imply that

$$dc_{it} = g(V_{it}, V_{i,t-1}, dw_{it}; \lambda) = dw_{it}(1 + \psi_0 V_{i,t-1}^{\alpha-1})/(1 + \psi_0 V_{it}^{\alpha-1}), \quad i = 1, 2$$

Given a vector λ and a realization of the exogenous state variables S_t – which includes money growth dM_{it} and endowment growth dw_{it} – consumption growth dc_{it} is an exact function of velocity, so the Euler equations above are fixed-point equations in the two velocity series. This means we can solve these first two Euler equations for the two equilibrium velocity processes as a unit before considering the remaining Euler equations. Using the solution velocity processes, we can then calculate directly equilibrium consumption growth dc_{it} and inflation dP_{it} for the two countries, and we can solve the remaining Euler equations to determine the equilibrium k-period interest rates in country 1, R_{1t}^k, the premium of the k-period forward rate over the spot rate f_t^k / e_t and exchange rate growth de_t.

Several methods have been used to solve nonlinear rational expectations models with endogenous state variables (Taylor and Uhlig, 1990; Judd, 1991). Among these, parameterized expectations is particularly suited to use with a simulation estimator because it produces simulated data that satisfy the Euler equations without having to solve for the full decision rule. We parameterize each of the conditional expectations in the above Euler equations as a function of the exogenous and endogenous state variables. The augmented vector of state variables is

$$\tilde{S}_t = (1, S_t, dc_{1,t-1}, dc_{2,t-1}, \cdots, dc_{1,t-L_c+1}, dc_{2,t-L_c+1}),$$

where 1 is concatenated for use as a constant in the regressions described below. If $L_c \le 1$, then there are no endogenous state variables, and $\tilde{S}_t$ is just equal to S_t and a constant. Any class of dense functions, such as polynomials or neural nets, can be used to approximate the conditional expectations. The particular functional form we use to parameterize expectations is

$$E_t(F_{ij,t}) = \exp[\text{poly}(\tilde{S}_t, \nu_{ij})],$$

where poly$(\cdot)$ is a polynomial in $\tilde{S}_t$, and ν_{ij} is the vector of its coefficients. We choose to use an exponential polynomial because economic theory implies that $E_t[F_{ij,t}]$ should be positive. In practice, the polynomial we use consists of linear and squared terms of the elements of $\tilde{S}_t$.

Below is a description of the algorithm for solving for the equilibrium velocity series given a vector λ. In every instance, the ranges of the indices are $i = 1, 2$ and $j = 1, 2$; superscripts indicate iteration numbers.

Step 1.
Simulate a realization of $\{u_t\}$, where u_t is *iid* $N(0, \Omega)$.

Step 2.
From some initial S_0, generate a realization of $\{S_t\}$ using

$$\log S_t = a_0 + A \log S_{t-1} + u_t.$$

In practice, we set S_0 to a vector of ones, but in performing the parameterized expectations regressions we exclude the first five hundred observations from the simulated data to eliminate any effect from choosing initial values.

Step 3.
Determine starting realizations of the velocity series $\{V_{it}^0\}$.

We consider two possible ways to do this. The first is to specify starting values for ν_{ij}^0, perhaps values of ν_{ij} obtained from a previous solution at a nearby λ. Then, given ν_{ij}^0 and some initial observations on velocities $V_{it}^0, t = 0, \ldots, L_c$, the remaining elements of the starting velocity series for $t = L_c + 1, \ldots, T$, can be determined using the following relationships recursively

$$dc_{it}^0 = g(V_{it}^0, V_{i,t-1}^0, dw_{it}; \lambda),$$

$$V_{it}^0 = \exp[\text{poly}(\tilde{S}_t^0, \nu_{i1}^0)] / \exp[\text{poly}(\tilde{S}_t^0, \nu_{i2}^0)].$$

This structure is recursive because $\tilde{S}_t^0$ contains $dc_{i,t-1}^0$. A drawback to this approach is that the simulated time series produced by the solution procedure are dependent upon the starting values, so any attempt to replicate the solution exactly would require knowing those starting values.

A second approach for establishing starting realizations of the velocity series would set V_{1t}^0 and V_{2t}^0 to be constants for all t. For these constants, one could calculate steady-state values for the two velocities, or simply set the velocities equal to 1. This latter approach still produces convergence in a relatively small number of iterations.

Regardless of the approach used to determine starting values of velocity, if one uses the procedure described below to improve the stability of the algorithm by dampening iteration updates, starting values must also be specified for the polynomial coefficients ν_{ij}^0. We recommend setting all of the coefficients to zero except the constants. This means that $E_t(F_{ij,t}) = \exp[\text{poly}(\tilde{S}_t, \nu_{ij})]$ reduces to $E_t(F_{ij,t}) = \exp[\text{constant}_{ij}]$. The constants can be set equal to the log of the unconditional means of the $F_{ij,t}$'s. Setting the initial polynomial coefficients in this way gives a very stable position from which to start the iterations.

Step 4.
Iteration k: Using the V_{it}^{k-1} series, calculate the $F_{ij,t}^{k-1}$ and regress each of these four on a linearized version of $\exp[\text{poly}(S_t{}^{k-1}, \nu_{ij}^k)]$ to estimate ν_{ij}^k. The linearization is done around ν_{ij}^{k-1}.

A linearized version of the exponential function is used to allow one to perform linear regressions rather than nonlinear regressions at each iteration. When the coefficients converge ($\nu_{ij}^k = \nu_{ij}^{k-1}$), the value of the exponential function is equal to the value of its linearized version at the point at which we want to evaluate it.

Den Haan and Marcet (1990) actually suggest a more gradual way of modifying the guesses at the polynomial coefficients from iteration to iteration. Rather than setting ν_{ij}^k equal to the coefficients obtained from the regressions, one can set ν_{ij}^k equal to a convex combination of those coefficients, call them b_{ij}^k, and the guess at the coefficients from the previous iteration as

$$\nu_{ij}^k = \rho b_{ij}^k + (1-\rho)\nu_{ij}^{k-1},$$

where $0 < \rho \leq 1$. This procedure has the effect of dampening the speed with which the guesses at the coefficients are updated. The smaller is ρ, the more gradually the coefficients are modified from one iteration to the next. One might want to use this gradual updating scheme to stabilize iterations that are not well behaved. For the model in this paper, we were always able to set $\rho = 1$, which implies no dampening in updating the coefficients.

Step 5.
Determine the two V_{it}^k series according to

$$V_{it}^k = \exp[\text{poly}(\tilde{S}_t^{k-1}, \nu_{i1}^k)] / \exp(\text{poly}[\tilde{S}_t^{k-1}, \nu_{i2}^k)],$$

and the two dc_{it}^k series according to

$$dc_{it}^k = g(V_{it}^k, V_{i,t-1}^k, dw_{it}; \lambda).$$

Step 6.
Repeat steps 3 and 4 until the velocity series converge. Convergence is reached when

$$\max_i \max_t |(V_{it}^k - V_{it}^{k-1})/(V_{it}^{k-1} + \epsilon)| \leq \xi,$$

where ϵ and ξ are small positive numbers.

Note that we check convergence on the velocity series, that is, on the ratios of the parameterized expectations projections, which is a different procedure than that used in Marcet (1991). Marcet looks for convergence of the coefficients of the projections, rather than of the projections themselves. We check convergence on the projections because of complications that arise when there is a high degree of multicollinearity between the variables of the parameterized polynomial, as is the case in our model. Multicollinearity makes it possible for the coefficients of the polynomial to continue to oscillate between successive iterations even though the projection onto the polynomial has essentially converged. Since it is the values of the projections that are important for solving the model, we look for convergence of those values.

In summary, the parameterized expectations solution method works by alternating between estimating values of conditional expectations based on some postulated realization of the velocity processes (which amounts to estimating the ν_{ij}'s) and updating the postulated values of the velocity processes based on the estimated conditional expectation values. The procedure continues until the velocity processes converge.

Once the equilibrium velocity and consumption growth series have been determined from the first two Euler equations, the four-period interest rate series in country 1, the premium of the four-period forward exchange rate over the spot rate, and the exchange rate growth can be determined from the remaining Euler equations without additional iterations. The Euler equations can be written as

$$R_{1t}^4 = \frac{E_t(F_{11,t})/(1+\alpha\psi_0 V_{1t}^{\alpha-1})}{\beta^4 E_t\left[f_{13}\left(dc_{1,t+1}, dc_{2,t+1}, \ldots, dc_{1,t+L_c+5}, dc_{2,t+L_c+5}, dP_{1,t+1}, \ldots, dP_{1,t+4}; \lambda\right)\right]},$$

$$e_t R_{1t}^4 / f_t^4 = \frac{E_t(F_{21,t})/(1+\alpha\psi_0 V_{2t}^{\alpha-1})}{\beta^4 E_t\left[f_{23}\left(dc_{1,t+1}, dc_{2,t+1}, \ldots, dc_{1,t+L_c+5}, dc_{2,t+L_c+5}, dP_{2,t+1}, \ldots, dP_{2,t+4}; \lambda\right)\right]},$$

$$de_t = \frac{E_t(F_{21,t})E_{t-1}(F_{11,t-1})(1+\alpha\psi_0 V_{1t}^{\alpha-1})(1+\alpha\psi_0 V_{2,t-1}^{\alpha-1})dP_{1t}dc_{1t}}{E_{t-1}(F_{21,t-1})E_t(F_{11,t})(1+\alpha\psi_0 V_{2t}^{\alpha-1})(1+\alpha\psi_0 V_{1,t1}^{\alpha-1})dP_{2t}dc_{2t}}.$$

In these equations $dP_{it} = (dM_{it}V_{it})/(dc_{it}V_{i,t-1})$, the gross inflation rate in each country. As before, f_{13} and f_{23} are particular function forms. The conditional expectations terms in the equations are each estimated by regressing the value of the function inside the expectations operator on a polynomial in $\tilde{S}_t$. The polynomial we use consists of the elements of $\tilde{S}_t$ raised to the first, second, third, and fourth powers. The resulting simulation values are used to form $\{y_t^\lambda\}$.

The time required to solve the structural economic model at some value of λ is an important consideration, since our nonparametric estimator requires solutions at many different values of λ in finding the value that minimizes the GMM objective function. When we use simulated time series of length 1000 to solve the model (excluding an initial discarded 500 observations), convergence for most values of λ is achieved in approximately one minute on a SUN SPARCstation 2.

3.2. *Defining the GMM Objective Function*

The Gallant-Tauchen (1992) SNP estimator underlies our nonparametric structural estimator. Following their notation, given the observed process $\{y_t\}$, let $x_{t-1} = (y'_{t-1}, \ldots, y'_{t-L})'$ and let $p(y_t|x_{t-1}, \lambda_0)$ denote the conditional density of y_t conditional on L lags of itself and the true λ_0. By stationarity, we can suppress the t subscript and simply write $p(y|x, \lambda_0)$ when convenient. In addition, let $p(y, x, \lambda_0)$ denote the joint density of (y_t, x_{t-1}). Frequently, we suppress the dependence of the conditional density on λ_0 and write $p(y|x)$, but we always make explicit the dependence of the joint density $p(y, x, \lambda_0)$ on λ_0, because that becomes important.

The SNP estimator is a sieve estimator that is based on the sequence of models $\{f_K(y|x, \theta_K)\}_{K=0}^{\infty}$, where $\theta_K \in \Theta_K \subseteq \Re^{\ell_K}$, $\Theta_K \subseteq \Theta_{K+1}$ and where $f(y|x, \theta_K)$ is a truncated Hermite series expansion. This hierarchy of models can, under regularity conditions, approximate $p(y|x)$ well in the sense

$$\lim_{K\to\infty} \inf_{\theta_K \in \Theta_K} \|[f_K(\cdot|\cdot, \theta_K) - p(\cdot|\cdot)\| = 0,$$

where $\|\cdot\|$ is a Sobelov norm. The approximation also holds along a sequence of estimated models fitted to data sets $\{y_{-L+1}, \ldots, y_n\}$, $n = 1, 2, \ldots \infty$, with the the appropriate model for each n determined by a model selection strategy.

The key component of our nonparametric structural estimator is the mean gradient of the log-density of a Kth order SNP model,

$$g(\lambda, \theta_K) = \int\int (\partial/\partial\theta) \log[f_K(y|x, \theta_K)] p(y, x, \lambda) dy dx.$$

In practice, the above expectation is approximated by simulating $\{U_\tau^\lambda\}_{\tau=1}^{\mathcal{T}}$, forming $\{y_\tau^\lambda\}$ as just described, taking lags to form $\{x_{\tau-1}^\lambda\}$, and then averaging

$$\tilde{g}(\lambda, \theta_K) = \frac{1}{\mathcal{T}} \sum_{\tau=1}^{\mathcal{T}} (\partial/\partial\theta) \log[f_K(y_\tau^\lambda | x_{\tau-1}^\lambda, \theta_K)].$$

We take $\tilde{g}(\lambda, \theta_K) \approx g(\lambda, \theta_K)$.

The nonparametric structural estimator is defined as follows: Let $\{y_t\}_{t=-L+1}^n$ be a realization of the observed process and let

$$\hat{\theta}_{Kn} = \arg \max_{\theta_K \epsilon \Theta_K} \left\{ \frac{1}{n} \sum_{t=1}^n \log[f(y_t | x_{t-1}, \theta_K)] \right\}.$$

Thus, $\hat{\theta}_{Kn}$ is the estimated parameter vector of a Kth order SNP model fitted to the data by maximum likelihood. The estimator $\hat{\lambda}$ is the solution of the GMM estimation problem

$$\hat{\lambda} = \arg \min_{\lambda \epsilon \Lambda} \{\tilde{s}_n(\lambda)\},$$

where

$$\tilde{s}_n(\lambda) = n \tilde{g}(\lambda, \hat{\theta}_{Kn})' W_n \tilde{g}(\lambda, \hat{\theta}_{Kn}),$$

and where W_n is a symmetric positive definite weighting matrix such that $W_n \to W$ almost surely and W is positive definite. In the application, we use

$$W_n = \left\{ \frac{1}{n} \sum_{t=1}^n (\partial/\partial\theta) \log[f(y_t | x_{t-1}, \hat{\theta}_{Kn})] (\partial/\partial\theta') \log[f(y_t | x_{t-1}, \hat{\theta}_{Kn}]) \right\}^{-1},$$

which is the natural estimate of the inverse of the information matrix based on the gradient-outer-product formula. This choice makes the minimized value of the GMM objective function, $\tilde{s}_n(\lambda)$, approximately $\mathcal{X}^2(\ell_K - \ell_\lambda)$ for large K.

Below we consider several different algorithms for minimizing $\tilde{s}_n(\lambda)$. Regardless of the algorithm, it is advantageous to control the interface between the optimizer and the economic model by scaling the optimizer's guesses at the parameter values to be within a range in accordance with the economic theory behind our model. For example, in our model it only makes sense for δ to be between 0 and 1, so we constrain the optimizer to attempt solutions only with such values. These constraints are imposed by using various forms of logistic transformations.

3.3. *Optimizing the Objective Function*

The basic computational task for the estimator is to evaluate $\hat{\lambda} = \text{argmin}_{\lambda \epsilon \Lambda} \{\tilde{s}_n(\lambda)\}$. This minimization is not straightforward for our problem because of the large number of parameters to be estimated (between 37 and 41 depending upon whether one, two, or three lags of consumption services enter the utility function) and because analytical derivatives of the objective function with respect to λ are not available. We tried four different algorithms for minimizing the objective function and found significant differences across algorithms for our problem.

3.3.1. *Optimizing with NPSOL and DFP*

We initially tried two classic gradient search methods: NPSOL (Gill, Murray, Saunders, and Wright, 1986), and Davidon-Fletcher-Powell (DFP), as implemented in the GQOPT package (Quandt and Goldfeld, 1991). Both algorithms work in a similar manner. A search direction is determined, a one-dimensional optimization is performed along that direction, and then the search direction is updated. The process is repeated until a putative optimum is achieved.

These algorithms work quite well when analytic derivatives are available. For example, we use NPSOL to perform the preliminary SNP parameter estimation to compute $\hat{\theta}_{Kn}$, which is needed to form $\tilde{s}_n(\lambda)$. Analytical derivatives are available for the SNP objective function, and NPSOL works adequately even on fairly large problems. In our application, the SNP estimation itself entails a specification search over roughly thirty different models with some having as many as 150 parameters. That whole effort takes only three or four days on a SUN SPARCstation 2. In an variety of other SNP applications, NPSOL has been found to work reasonably well (Gallant and Tauchen, 1992).

Analytical derivatives of $\tilde{s}_n(\lambda)$, however, are computationally infeasible. The process $\{y_t^\lambda\}$ is a solution to a fixed-point problem, as are its analytical derivatives. Computing $\partial\tilde{s}_n(\lambda)/\partial\lambda$ would involve computing a solution to a fixed-point problem for each component. Evaluating $\tilde{s}_n(\lambda)$ and its derivatives for arbitrary λ is well beyond the reach of current computing equipment. The large computational demands for analytical derivatives appear to be intrinsic to all solution methods for nonlinear structural models, including those described in the *JBES* Symposium (Tauchen, 1990; Taylor and Uhlig, 1990) or Judd (1991), since they all entail solving nonlinear fixed-point problems.

Gradient search methods use numerical derivatives in place of analytical derivatives when the latter are unavailable. For our type of problem, this does not work well. The computations turn out to be about as demanding as would be those for analytical derivatives approximating the gradient of the objective function $\partial \tilde{s}_n(\lambda)/\partial\lambda$ at a single point λ entails computing the simulated process $\{y_t^\lambda\}$ after small perturbations in each of the λ. With ℓ_λ on the order of 37 to 41, this entails, at a minimum, recomputing the equilibrium of the model that many additional times just to approximate a single one-sided gradient. The net effect is to generate about as many function calls as would a naive grid-search. In fact, our experience suggests that a naive grid search might even work better. In the course of approximating $\partial \tilde{s}_n(\lambda)/\partial\lambda$ via perturbing λ and forming difference quotients, values of λ that produce sharp improvement in the objective function are uncovered quite by happenstance. Neither NPSOL nor DFP retains and makes use subsequently of these particularly promising values of λ; the effort that goes into to computing the equilibrium for these λ is lost. Simple grid search would retain these λ's.

3.3.2. *Optimizing with Simulated Annealing*

We also tried simulated annealing, a global method. An implementation of simulated annealing by William Goffe is available in the GQOPT optimization package (Quandt and Goldfeld, 1991). We used an updated version that William Goffe kindly made available to us. See Goffe, Ferrier, and Rodgers (1992) for a discussion of the algorithm and additional references. We give a brief summary of the essential ideas here.

From a point λ, simulated annealing changes element i of λ using

$$\lambda_i' = \lambda_i + rv_i,$$

where r is a uniformly distributed random number over $[-1, 1]$ and v_i is the ith element of a vector of weights V. If $\tilde{s}_n(\lambda')$ is smaller than $\tilde{s}_n(\lambda)$ the point is accepted. If not, the point is accepted if a random draw from the uniform over [0, 1] exceeds

$$p = e^{[\tilde{s}_n(\lambda') - \tilde{s}_n(\lambda)]/T}.$$

The elements of V and T are tuning parameters that must be selected in advance and are adjusted throughout the course of the iterations. We used the defaults. There are additional tuning parameters that determine when these adjustments occur. Again, we accepted the defaults.

The algorithm was defeated by the large number of function evaluations that it requires. Most exasperating was its insistence on exploring unprofitable parameter values. After making some promising initial progress the algorithm would plateau far from an optimum and give no indication that further progress could be achieved if the iterations were permitted to continue.

3.3.3. *Optimizing with Simplex Methods*

The optimization method that performs best for our problem is the simplex method developed by Nelder and Mead (1964). Fortran code for implementing this method is available in the GQOPT optimization package (Quandt and Goldfeld, 1991). The method works as follows: We begin the minimization of a function of ℓ_λ variables by constructing a simplex of $(\ell_\lambda + 1)$ points in ℓ_λ-dimensional space: λ_0, λ_1, ..., λ_{ℓ_λ}. We denote the value of the function at point λ_i by s_i. The lowest, highest, and second highest values are

$$s_l = \min_i(s_i), \qquad s_h = \max_i(s_i), \qquad s_{hh} = \max_{i \neq h}(s_i),$$

corresponding to points λ_l, λ_h, and λ_{hh}. We also define the notation $[\lambda_i \lambda_j]$ to indicate the distance from λ_i to λ_j.

The algorithm works by replacing λ_h in the simplex continuously by another point with a lower function value. Three operations are used to search for such a new point–*reflection*, *contraction*, and *expansion*–each of which is undertaken relative to the centroid $\bar{\lambda}$ of the simplex points excluding λ_h. The centroid is constructed as

$$\bar{\lambda} = \frac{1}{\ell_\lambda} \sum_{i=1}^{\ell_\lambda} \lambda_i, \qquad i \neq h.$$

The *reflection* of λ_h through the centroid is λ_r, which is defined by

$$\lambda_r = (1 + \alpha_r)\bar{\lambda} - \alpha_r \lambda_h,$$

where $\alpha_r > 0$ is the *reflection coefficient*. λ_r lies on the line between λ_h and $\bar{\lambda}$, on the far side of $\bar{\lambda}$, and α_r is the ratio of the distance $[\lambda_r \bar{\lambda}]$ to $[\lambda_h \bar{\lambda}]$. If $s_l < s_r \leq s_{hh}$, we replace λ_h with λ_r and start the process again with this new simplex.

If reflection has produced a new minimum $(s_r < s_l)$, we search for an even lower function value by expanding the reflection. The *expansion* point is defined by

$$\lambda_e = \alpha_e \lambda_r + (1 - \alpha_e)\bar{\lambda}$$

where $\alpha_e > 1$ is the *expansion coefficient* that defines the ratio of the distance $[\lambda_e \bar{\lambda}]$ to $[\lambda_r \bar{\lambda}]$. λ_e is farther out than λ_r on the line between λ_h and $\bar{\lambda}$. If $s_e < s_r$, λ_h is replaced in the simplex by λ_e. Otherwise, the expansion has failed and λ_r replaces λ_h. The process is then restarted with the new simplex.

If reflection of λ_h has not even produced a function value less than s_{hh} – which means that replacing λ_h with λ_r would leave s_r the maximum – we rename λ_h to be either the old λ_h or λ_r, whichever has a lower function value. Then we attempt to find an improved point by constructing the *contraction*

$$\lambda_c = \alpha_c \lambda_h + (1 - \alpha_c)\bar{\lambda},$$

where $0 < \alpha_c < 1$. The *contraction coefficient* α_c is the ratio of the distance $[\lambda_c \bar{\lambda}]$ to $[\lambda_h \bar{\lambda}]$. If $s_c < s_h$, then the contraction has succeeded, and we replace λ_h with λ_c and restart the process. If this contraction has failed, we construct a new simplex by contracting all the points toward the one with the lowest function value, which is accomplished by replacing the λ_i's with $(\lambda_i + \lambda_l)/2$. Then the process of updating the simplex restarts.

Nelder and Mead suggest stopping their procedure when the standard deviation of the λ_i's is less than some critical value. In our empirical work, we strengthen this stopping rule by restarting the algorithm several times from the value on which the Nelder-Mead procedure settles. When this restarting leads to no further significant improvement in the objective function value, we accept the best point as the minimum of the function. In implementing the algorithm, we also found it advantageous to modify the error handling procedures of the Nelder-Mead code provided in GQOPT slightly to allow us to start the procedure with a wider ranging simplex.

The Nelder-Mead simplex method was far more successful than the other methods we tried for minimizing our objective function. There are two aspects of this method that we believe are responsible for its success. First, the method finds new lower points on the objective surface without estimating derivatives. Second, by using the operations of *reflection*, *expansion*, and *contraction*, the Nelder-Mead method is designed to jump over ridges in the objective surface easily in searching for new lower points. This property can be important in preventing an optimization algorithm from shutting down too early. Despite these advantages, however, the performance of the Nelder-Mead method is not completely satisfactory, because it requires a very large number of function calls to find the minimum of the function. Given the number of parameters in our model and the complexity of evaluating the objective function at any one point, the method can occupy several weeks of computing time on a Sun SPARCstation. Even though this computing demand is substantial and far greater than we expected from the outset of this project, we still consider our nonparametric structural estimator very successful in achieving our goal of estimating a nonlinear rational expectations model and fully accounting for the complex nonlinear dynamics of actual time series in that estimation.

Results from applying this estimator to the illustrative monetary model are available in Bansal, Gallant, Hussey, and Tauchen (1992).

4. CONCLUSION

In this paper we describe a new nonparametric estimator for structural equilibrium models and show its application to an equilibrium monetary model. The discussion of the implementation of the estimator indicates important considerations that might arise in applying the estimator to other nonlinear rational expectations models.

There are several advantages to this estimator. By using the method of parameterized expectations to solve the model numerically, structural equilibrium models can be estimated without limiting oneself to linear approximations. By using a consistent nonparametric estimate of the conditional density of the observed data to define the criterion to be minimized in estimation, the estimator forces the model to confront the law of motion of the observed data, which can include complex forms of nonlinearity. Finally, the estimator provides simulated data from the model. If a model is rejected, then it is possible to evaluate the dimensions in which it fails to match characteristics of the observed data, thus providing valuable diagnostic information for building better models.

ACKNOWLEDGEMENTS

This material is based upon work supported by the National Science Foundation under Grants No. SES-8808015 and SES-90-23083. We thank Geert Bekaert, Lars Hansen, David Hsieh, Ellen McGrattan, Tom Sargent, and many seminar and conference participants for helpful comments at various stages of this research.

REFERENCES

Bansal, R., 1990, "Can non-separabilities explain exchange rate movements and risk premia?", Carnegie Mellon University, Ph.D. dissertation.

Bansal, R., A. R. Gallant, R. Hussey and G. Tauchen, 1992, "Nonparametric estimation of structural models for high-frequency currency market data", Duke University, manuscript.

Bollerslev, T., 1986, "Generalized autoregressive conditional heteroskedasticity", *Journal of Econometrics* 31, 307–327.

den Haan, W. J. and A. Marcet, 1990, "Solving the stochastic growth model by parameterizing expectations", *Journal of Business and Economic Statistics* 8, 31–4.

Duffie, D. and K. J. Singleton, 1989, "Simulated moments estimation of markov models of asset prices", Stanford University, Graduate School of Business, manuscript.

Dunn, Kenneth and K. J. Singleton, 1986, "Modeling the term structure of interest rates under non-separable utility and durability of goods", *Journal of Financial Economics* 17, 27–55.

Engle, R. F., 1982, "Autoregressive conditional heteroscedasticity with estimates of the variance of United Kingdom inflation", *Econometrica* 50, 987–1007.

Feenstra, R. C., 1986, "Functional equivalence between liquidity costs and the utility of money", *Journal of Monetary Economics* 17, 271–291.

Gallant, A. R. and G. Tauchen, 1989, "Seminonparametric estimation of conditionally constrained heterogeneous processes: asset pricing applications", *Econometrica* 57, 1091–1120.

Gallant, A. R. and G. Tauchen, 1992, "A nonparametric approach to nonlinear time series: estimation and simulation", in David Brillinger, Peter Caines, John Zeweke, Emanuel Paryen, Murray Rosenblatt, and Murad S. Taggu (eds.), *New Directions in Time Series Analysis, Part II*, New York: Springer-Verlag, 71–92.

Gill, P. E., W. Murray, M. A. Saunders and M. H. Wright, 1986, "User's guide for NPSOL (version 4.0): a Fortran package for nonlinear programming", Technical Report SOL 86-2, Palo Alto: Systems Optimization Laboratory, Stanford University.

Goffe, W. L., G. D. Ferrier, and J. Rodgers, 1992, "Global Optimization of statistical functions: Preliminary results" in Hans M. Amman, David A. Belsley, and Louis F. Pau (eds.), *Computational Economics and Econometrics, Advanced Studies in Theoretical and Applied Econometrics*, Vol. 22, 19–32, Boston: Kluwer Academic Publishers.

Hansen, L. P., 1982, "Large sample properties of generalized method of moments estimators", *Econometrica* 50, 1029–1054.

Hansen, L. P. and T. J. Sargent, 1980, "Formulation and estimation of dynamic linear rational expectations models", *Journal of Economic Dynamics and Control* 2, 7–46.

Hansen, L. P. and K. J. Singleton, 1982, "Generalized instrumental variables estimators of nonlinear rational expectations models", *Econometrica* 50, 1269–1286.

Ingram, B. F. and B. S. Lee, 1991, "Simulation estimation of time-series models", *Journal of Econometrics* 47, 197–205.

Judd, K. L., 1991, "Minimum weighted least residual methods for solving aggregate growth models", Federal Reserve Bank of Minneapolis, Institute of Empirical Macroeconomics, manuscript.

Lucas, R. E., Jr., 1982, "Interest rates and currency prices in a two-country world", *Journal of Monetary Economics* 10, 335–360.

Marcet, A., 1991, "Solution of nonlinear models by parameterizing expectations: an application to asset pricing with production", manuscript.

McCallum, B. T., 1983, "On non-uniqueness in rational expectations models: an attempt at perspective", *Journal of Monetary Economics* 11, 139–168.

Nelder, J. A. and R. Mead, 1964, "A simplex method for function minimization", *The Computer Journal* 7, 308–313.

Quandt, R. E. and S. M. Goldfeld, 1991, *GQOPT/PC*, Princeton, N.J.

Stigum, M., 1990, *The money market*, 3rd ed., Homewood, Il.: Dow Jones–Irwin.

Svensson, L. E. O., 1985, "Currency prices, terms of trade and interest rates: a general equilibrium asset-pricing cash-in-advance approach", *Journal of International Economics* 18, 17–41.

Tauchen, G., 1990, "Associate editor's introduction", *Journal of Business and Economic Statistics* 8, 1.

Taylor, J. B. and H. Uhlig, 1990, "Solving nonlinear stochastic growth models: a comparison of alternative solution methods", *Journal of Business and Economic Statistics* 8, 1–17.

A.J. HUGHES HALLETT AND YUE MA

On the Accuracy and Efficiency of GMM Estimators: A Monte Carlo Study

ABSTRACT. GMM estimators are now widely used in econometric and financial analysis. Their asymptotic properties are well known, but we have little knowledge of their small sample properties or their rate of convergence to their limiting distribution. This paper reports small sample Monte Carlo evidence which helps discriminate between the many GMM estimators proposed in the literature. We add a new GMM estimator which delivers better finite sample properties. We also test whether biases in the parameter estimates are either significant or significantly different between estimators. We conclude that they are, with both relative and absolute biases depending on sample size, fitting criterion, non-normality of disturbances, and parameter size.

1. INTRODUCTION

One of the most interesting developments in econometric theory over the past decade has been the introduction of the General Method of Moments (GMM) estimators. Not only is this a significant development because it offers a new and more flexible approach to estimation, it also opens up an estimation methodology that is particularly well suited to a range of problems – such as the econometrics of financial markets – where the form of the probability distributions, as well as their parameters, plays an important role.

The theoretical properties of GMM estimators – consistency, asymptotic efficiency and sufficiency – were established rapidly after Hansen first introduced the concept (Hansen, 1982). These properties are established in Duffie and Singleton (1989), Smith and Spencer (1991) and Deaton and Laroque (1992). However, few results have been presented on the GMM's small sample properties or rate of convergence to consistency. This would provide important information on the general reliability of GMM estimators. It seems that we lack such information because, although the principle of GMM estimation is well defined, there is no obvious agreement on the algorithms to be used for computing the estimates themselves. The theoretical contributions have been vague on implementation and the choice of fitting criterion. This paper examines 7 different suggestions from the recent literature.

The first purpose of this paper is to provide some empirical experience that helps the user discriminate between different GMM estimation techniques. Second, we introduce a new GMM estimator which, in our experiments at least, produces better finite sample results than any of the other techniques reported in the literature. Third,

D. A. Belsley (ed.), Computational Techniques for Econometrics and Economic Analysis, 23–44.

we draw a distinction between the case where we have to estimate a few model parameters conditionally on an assumed distribution for the random components (the traditional econometric approach) and the more general problem of fitting a whole distribution or probability model.

2. THE GMM ESTIMATORS STUDIED

Most GMM estimators can be specified within the framework established by Hansen (1982). That framework exploits the general orthogonality condition

$$Eg(x_t, \beta) = 0\,, \tag{1}$$

where β is a k-vector of parameters, x_t is a T-vector of data, and $g(x, \beta)$ is an m-vector of functions of data and parameters. We would have (1) if the maintained hypothesis is a conventional econometric model, say, $y_t - h(X_t, \beta) = u_t$. That gives the standard regression approach, where we try to minimise some function of $g(\cdot) = y_t - h(x_t, \hat{\beta})$ as a sample of T observations taken right through the distribution of u_t. But we could also pick $\hat{\beta}$ to make the *fitted* u_t distribution yield the same characteristics as we actually observe in the data on y_t, given x_t. More generally, we pick $\hat{\beta}$ to minimise

$$g(x_t, \beta) = T^{-1} \sum [f(y_t) - \hat{f}(x_t, \hat{\beta})]\,, \tag{2}$$

where $f(y_t)$ is any function of the observed data, and $\hat{f}(x_t, \hat{\beta})$ is its fitted counterpart under the maintained hypothesis and chosen parameter values $\hat{\beta}$.

In practice we have to define the best fit in some metric, i.e. we choose $\hat{\beta}$ by solving

$$\min_{\beta} \{J(g(x_t, \beta)) = \|g(\cdot)\|_{r,w}\}\,, \tag{3}$$

where the value of r defines the norm and W the weighting function. This is the GMM strategy when the $f(\cdot)$ represent a series of moments from the probability distribution of y_t; that is, $f(\cdot)$ defines the sample moments, and $\hat{f}(\cdot)$ represents the fitted moments given x_t and the choice of $\hat{\beta}$. In many cases we do not have an analytic maintained hypothesis, so the fitted moments $\hat{f}(\cdot)$ have to be constructed by numerical simulation with pseudo-data replicated many times through the model to generate numerical evaluations of those moments. That variant is the method of simulated moments.

Now, if (1) is correct the sample moment,

$$g_T(\beta) = \sum_{t=1}^{T} g(x_t, \beta)\,/\,T$$

should be close to zero when evaluated at $\beta = \hat{\beta}$. It is therefore reasonable to estimate β by choosing $\hat{\beta}$ to minimise

$$J_T(\beta) = g_T(\beta)'\, W_T g_T(\beta)\,,$$

where W_T is a positive definitive weighting matrix. Setting $W_T = I$ or Ω^{-1}, where $Euu' = \Omega$, gives the OLS or GLS regression based GMM estimators, and setting $W_T = N(N'N)^{-1}N'$, gives the instrumental variable version with instruments N. Varying the specification of $g(\cdot,\cdot)$ gives different GMM estimators:

(1) The simple method of moments.
Define $g(x_t, \beta) = [x_t - \mu_1,\ \frac{T}{T-1}\,(x_t - m_1)^2 - \mu_2]'$ *and* $W_T = I_2$. *Then*

$$g_T(\beta) = \left[\sum x_t/T - \mu_1,\ \sum (x_t - m_1)^2\,/\,(T-1) - \mu_2\right]'$$

$$= [m_1 - \mu_1,\ m_2 - \mu_2]'\,,$$

where m_i is the i-th sample moment and μ_i is the corresponding central moment from the probability density function expressed in terms of the parameters of the underlying theoretical model. The simplest GMM estimator minimises

$$J_T(\beta) = g_T(\beta)' \cdot g_T(\beta) = \sum_{j=1}^{2} (m_j - \mu_j)^2\,.$$

The solution to this problem is to set $\hat{\mu}_1 = m_1$ and $\hat{\mu}_2 = m_2$.

(2) The method of simulated moments (Smith and Spencer, 1991)

Define

$$g(x_t, \beta) = \left[x_t - \mu_1,\ \frac{T}{T-1}\,(x_t - m_1)^2 - \mu_2,\right.$$
$$\left.\frac{T}{T-1}\,(x_t - m_1)^3 - \mu_3,\ \frac{T}{T-1}\,(x_t - m_1)^4 - \mu_4\right]'$$

or

$$\tilde{g}(x_t, \beta) = \left[x_t - \mu_1,\ \frac{T}{T-1}\,(x_t - m_1)^2 - \mu_2,\ \frac{T}{T-1}\,(x_t - m_1)^3 - \mu_3\right]'.$$

The Duffie and Singleton (1989) GMM estimator then minimises

$$J_{4T}(\beta) = g_T(\beta)'\, g_T(\beta) = \sum_{j=1}^{4} (m_j - \mu_j)^2\,,$$

while Smith and Spencer's (1991) version considers only the first three moments

$$J_{3T}(\beta) = \tilde{g}_T(\beta)'\, \tilde{g}_T(\beta) = \sum_{j=1}^{3} (m_j - \mu_j)^2\,.$$

(3) Our new GMM method (Hughes Hallett, 1992):

Define $g_{4T}(\beta) = \Big[\sum x_t/T - \mu_1,\ \sqrt{\sum (x_t - m_1)^2/(T-1)} - \sqrt{\mu_2}\,,$

$$\sqrt[3]{\sum (x_t - m_1)^3/(T-1)} - \sqrt[3]{\mu_3}\,,$$

$$\sqrt[4]{\sum (x_t - m_1)^4/(T-1)} - \sqrt[4]{\mu_4}\Big]'\,,$$

and then construct a GMM estimator by minimising

$$J_T(\beta) = \sum_{j=1}^{4} (m_j^{1/j} - \mu_j^{1/j})^2\,.$$

This estimator has the advantage that each element in the objective function is expressed in the same units of measurement. It therefore produces a result that is independent of those units, and hence of the weighting of the moments implicit in the previous two estimators.

(4) The Newey and West (1987) version of the GMM estimator introduces the weighting matrix $W_T = V_T^{-1}$, where

$$V_T = \sum_{t=1}^{T} g(x_t, \beta*)\, g'(x_t, \beta*)/T\,,$$

and where $\beta*$ is an initial estimator of β obtained from an unweighted method of simulated moments estimator[1].

(5) The Deaton-Laroque method (Deaton and Laroque, 1992) is an instrumental variable based GMM estimator, adapted here to the problem of fitting a complete distribution. Define $Z_t = (1, x_{t-1}, x_{t-2}, x_{t-3})'$ and $Z = (Z_1', Z_2', \ldots, Z_T')'$, where the x_t's are actual observations in ascending order $t = 1 \ldots T$.

Now let

$$a_{t-1} = \frac{x_t + x_{t-1}}{2}\,,\quad a_0 = 0\,,\quad a_T = +\infty\,,$$

and $f(x, \beta)$ is the probability density function of x_t. Then u_t represents the difference between the actual frequency of observations in the interval $[a_{t-1}, a_t]$ and the theoretical probability of getting an observation in the same interval according to the maintained hypothesis. We then pick β to minimise the "error" between the two probabilities, that is, to maximise the fit between the observed relative frequencies and the maintained probability. To do that, write $u = (u_1, u_2, \ldots, u_T)'$, together with

[1] Other values are obviously possible for β*, and it would certainly be possible to iterate on the β* values to give a fully converged GLS style estimator.

$$g(x_t, \beta) = T \cdot Z_t' u_t \,, \quad \text{and} \quad g_T(\beta) = \sum_{t=1}^{T} g(x_t, \beta)/T = Z' u \,.$$

Let $W_T = (Z'Z)^{-1}$. Then the Deaton-Laroque estimator minimises

$$J_T(\beta) = g_T' \, W_T \, g_T = u' \, Z(Z'Z)^{-1} \, Z' u \,.$$

We have followed Tauchen (1986) and Deaton and Laroque in using lagged observations from the data as instruments. But ordering those observations at discrete intervals over the data's range is necessary to construct the "data" sets for our Monte Carlo experiments.

(6) Finally, we also compare the GMM estimators with the Maximum likelihood (ML) estimator. Suppose the pdf is $f(x_t, \beta)$. The ML estimator maximises

$$J(\beta) = \prod_{t=1}^{T} f(x_t, \beta) \,.$$

(7) The existing literature sheds very little light on the small sample properties of GMM estimators. There appear to be just two studies that attempt to do so. Tauchen (1986) and Gregory and Smith (1990) both look at the performance of GMM estimators in a very particular model of assets in macroeconomic performance. Gregory and Smith find (i) that small sample bias increases as the size of the parameter being estimated increases (as we do below), (ii) the confidence intervals on the GMM estimates shrink significantly with increasing sample size (as we do), and (iii) the rate of convergence to consistency slows noticeably when there are stronger dynamics in the model. Tauchen concentrates on the instrumental variable version of the GMM estimator in the same model and finds that the finite sample results are not sensitive to the choice of instruments. But all of this is done with just 2 sample sizes, 2 GMM estimators, 2 parameter settings, and 1 maintained hypothesis.

Deaton and Laroque (1992), by contrast, use a different model and report poor estimates of the underlying distributions with small *and* medium sample sizes. No details are given; but evidently rather large samples may be needed to achieve the desired asymptotic properties – depending on the model, parameter values, estimation criterion, and distributional context chosen. That calls for closer investigation, both between different GMM estimators and relative to traditional estimation techniques.

3. THE MONTE CARLO EXPERIMENTS

To compare the performances of the various GMM estimators and the ML estimator in finite samples, we ran a series of Monte Carlo experiments using sample sizes of 20 and 200 and a variety of parameters values. First, we took 5 cases of the normal distribution with parameter values $(\mu, \sigma^2) = \{(0,1), (0,0.25), (0,2), (2,0.25), (2,2)\}$; then 3 cases of the gamma distribution with parameters $(\tau, \lambda) = \{(1,3),$

TABLE 1

Normal dist. SAMPLE SIZE = 200	1st param. μ		2nd param. σ^2		average χ^2 (d.f. = 17)	p-value
METHOD	bias	variance	bias	variance		
TRUE PARAM. $(\mu, \sigma^2) = (0, 1)$						
AHH	0.00430	0.00484	−0.01135	0.01207	11.1	85%
HNW	0.00467	0.00617	−0.05450	0.01225	11.7	
D–L	0.00471	0.00941	−0.07271	0.01385	12.2	
DS	0.00480	0.00984	−0.07334	0.01388	12.6	
SIMPLE/SS(3)/ML	0.00492	0.00985	−0.07675	0.01390	13.5	70%
TRUE PARAM. $(\mu, \sigma^2) = (0, 2)$						
AHH	0.00611	0.00970	−0.02309	0.04829	13.2	72%
HNW	0.00717	0.01233	−0.10901	0.04900	14.3	
D–L	0.00772	0.01879	−0.20413	0.05466	15.1	
DS	0.00780	0.01949	−0.22721	0.05532	15.6	
SIMPLE/SS(3)/ML	0.00790	0.01970	−0.23351	0.05678	16.0	52%
TRUE PARAM. $(\mu, \sigma^2) = (0, 1/4)$						
AHH	0.00220	0.00119	−0.00283	0.00074	10.7	89%
HNW	0.00253	0.00152	−0.01360	0.00076	12.1	
D–L	0.00322	0.00235	0.04903	0.00087	13.0	
DS	0.00415	0.00241	0.09736	0.00090	13.4	
SIMPLE/SS(3)/ML	0.00425	0.00251	0.09831	13.7	69%	
TRUE PARAM. $(\mu, \sigma^2) = (2, 2)$						
AHH	0.00608	0.00968	−0.02301	0.04817	12.1	79%
HNW	0.00731	0.01033	−0.10812	0.04893	12.5	
D–L	0.01150	0.01994	−0.61343	0.34311	16.0	
DS	0.00802	0.01908	−0.11001	0.05521	13.2	
SIMPLE/SS(3)/ML	0.00808	0.01968	−0.11351	0.05663	13.3	71%
TRUE PARAM. $(\mu, \sigma^2) = (2, 1/4)$						
AHH	0.00215	0.00121	−0.01254	0.00075	10.5	89%
HNW	0.00261	0.00156	−0.01382	0.00077	10.7	
D–L	0.01282	0.00164	0.01715	0.00303	11.0	
DS	0.00255	0.00171	−0.02164	0.00078	11.5	
SIMPLE/SS(3)/ML	0.00256	0.00172	−0.02269	0.00069	11.7	80%

$(3, 1), (1, 1)\}$; and finally 3 examples of the beta distribution with $(p, q) = \{(1, 3), (3, 1), (1, 1)\}$. In each experiment, 500 estimation replications were carried out. The NAG Library Fortran subroutines were used to generate the random pseudo data,

Table 1 (continued)

SAMPLE SIZE = 20						
METHOD	bias	variance	bias	variance	(d.f. = 4)	
TRUE PARAM. $(\mu, \sigma^2) = (0, 1)$						
AHH	0.01103	0.04674	–0.13735	0.09080	3.8	43%
HNW	0.01298	0.05960	–0.18649	0.09332	4.2	
D–L	–0.01365	0.46749	0.19824	2.65313	5.2	
DS	0.01402	0.06273	–0.23965	0.94469	6.1	
SIMPLE/SS(3)/ML	0.01403	0.06374	–0.29827	0.10405	6.2	20%
TRUE PARAM. $(\mu, \sigma^2) = (0, 2)$						
AHH	0.01885	0.09546	–0.25392	0.36118	4.5	34%
HNW	0.01938	0.12032	–0.57394	0.37404	5.1	
D–L	–0.02874	0.28734	0.79375	1.31727	5.8	
DS	0.01985	0.19526	–0.88349	0.37663	6.0	
SIMPLE/SS(3)/ML	0.01995	0.19550	–0.89653	0.42219	6.8	16%
TRUE PARAM. $(\mu, \sigma^2) = (0, 1/4)$						
AHH	0.00502	0.01201	–0.03184	0.00612	4.0	40%
HNW	0.00658	0.01501	–0.07042	0.00774	4.5	
D–L	–0.04946	0.50263	0.07546	1.63030	6.1	
DS	0.00702	0.01586	–0.08041	0.00816	5.3	
SIMPLE/SS(3)/ML	0.00712	001602	–0.09457	0.00888	5.5	24%
TRUE PARAM. $(\mu, \sigma^2) = (2, 2)$						
AHH	0.01787	0.09448	–0.25470	0.35120	3.5	48%
HNW	0.01838	0.11925	–0.27394	0.37414	4.1	
D–L	0.01935	0.12802	0.27443	0.52655	5.1	
DS	0.01979	0.19548	–0.28349	0.42263	5.5	
SIMPLE/SS(3)/ML	0.01992	0.19633	–0.29653	0.47619	5.7	23%
TRUE PARAM. $(\mu, \sigma^2) = (2, 1/4)$						
AHH	0.00602	0.01194	–0.03194	0.00511	3.1	54%
HNW	0.00640	0.01485	–0.07131	0.00574	4.2	
D–L	0.09401	0.05068	0.31639	2.84743	6.7	
DS	0.00707	0.01567	–0.12041	0.00616	5.1	
SIMPLE/SS(3)/ML	0.00713	0.01594	–0.13457	0.00688	5.8	22%

and each of the 7 estimators were applied to the resulting 500 "data" sets. In this, the start-up seeds were randomised by the clock and a quasi-Newton algorithm was used to find a minimum of a non-linear function subject to fixed upper and lower bounds for the range of possible parameter values. For the estimates themselves, the selection criteria are (1) unbiasedness

TABLE 2

Gamma dist. SAMPLE SIZE = 200	1st param. τ		2nd param. λ		average χ^2 p-value	
METHOD	bias	variance	bias	variance	(d.f. = 17)	
TRUE PARAM. $(\tau, \lambda) = (3, 1)$						
AHH	0.04330	0.08775	0.01061	0.01175	14.4	64%
HNW	0.05774	0.13113	0.01534	0.01627	16.2	
D–L	−0.14282	0.19760	−0.05869	0.02415	17.1	
DS	0.17742	0.57358	0.06290	0.04682	17.8	
SS(3)	0.22081	0.56552	0.05947	0.04773	18.7	
ML	0.18372	0.31388	0.05623	0.03518	18.0	40%
SIMPLE	0.33591	0.28049	0.11805	0.03183	19.5	
TRUE PARAM. $(\tau, \lambda) = (1, 3)$						
AHH	0.01474	0.00781	0.03886	0.12880	13.4	70%
HNW	0.02901	0.01783	0.08096	0.23007	14.0	
D–L	0.04951	0.03073	0.14412	0.34351	16.1	
DS	−0.04837	0.04014	−0.18473	0.42035	16.4	
SS(3)	0.08781	0.04403	0.25613	0.45788	17.0	
ML	0.13395	0.03310	0.48601	0.39696	20.1	37%
SIMPLE	0.09983	0.05107	0.28132	0.51546	18.5	
TRUE PARAM. $(\tau, \lambda) = (1, 1)$						
AHH	0.01468	0.00778	0.01295	0.01431	10.5	89%
HNW	0.02821	0.01781	0.02699	0.02556	11.4	
D–L	−0.04841	0.04018	−0.06160	0.04678	13.0	
DS	0.08486	0.05814	0.07379	0.05522	15.2	
SS(3)	0.09801	0.05212	0.09377	0.05727	15.8	
ML	0.14035	0.03619	0.17030	0.04789	16.4	49%
SIMPLE	0.13023	0.08550	0.11935	0.07498	17.3	

$$bias = \frac{1}{N} \sum_{i=1}^{N} (\hat{\beta}_i - \beta) \, ,$$

where β represents the true parameter value; and N is the number of replications ($N = 500$) and (2) efficiency

$$variance = \frac{1}{N-1} \sum_{i=1}^{N} (\hat{\beta}_i - \bar{\beta})^2 \, , \quad where \ \bar{\beta} = \frac{1}{N} \sum_{i=1}^{N} \hat{\beta}_i \, .$$

The choice of Normal, Gamma, and Beta densities in these experiments covers the wide variety of distribution shapes that are found in economic and financial data.

Table 2 (continued)

SAMPLE SIZE = 20						
TRUE PARAM. $(\tau, \lambda) = (3, 1)$					(d.f. = 4)	
AHH	0.55093	1.67360	0.18058	0.22084	5.2	27%
HNW	0.65978	2.00182	0.21690	0.25896	5.9	
D–L	0.77432	8.77511	0.20008	1.42427	6.4	
DS	1.52956	5.66632	0.48944	0.50706	6.7	
SS(3)	1.84638	7.06289	0.55156	0.71606	7.0	
ML	1.60091	6.13546	0.51914	0.69212	6.8	8%
SIMPLE	2.28789	6.09199	0.77963	0.68508	7.3	
TRUE PARAM. $(\tau, \lambda) = (1, 3)$						
AHH	0.13228	0.11975	0.59298	2.12712	5.9	21%
HNW	0.27328	0.2228	1.04345	3.40449	6.2	
D–L	0.33759	0.93460	1.02650	9.76649	6.4	
DS	0.38921	0.26057	1.37600	3.69683	6.9	
SS(3)	0.50374	0.27059	1.69972	3.68184	7.0	
ML	0.69007	0.58224	2.28504	7.90659	7.3	7%
SIMPLE	0.89671	0.67855	3.06943	7.95057	7.5	
TRUE PARAM. $(\tau, \lambda) = (1, 1)$						
AHH	0.13910	0.12316	0.20486	0.23759	3.8	43%
HNW	0.28118	0.22167	0.35581	0.37290	4.1	
D–L	0.64287	0.47650	0.66815	0.63571	4.7	
DS	0.70516	0.62484	0.77703	0.93212	6.1	
SS(3)	0.82113	0.49703	0.85024	0.75024	6.7	
ML	0.71317	6.03977	0.71959	5.88462	6.5	9%
SIMPLE	1.03394	0.92628	1.20932	1.27444	7.3	

4. RESULTS: PARAMETER ESTIMATION

Tables 1 to 3 contain the results of our Monte Carlo parameter estimation experiments for the Normal, Gamma, and Beta distribution cases, respectively. For each estimate of them we report the bias and variance achieved by our 7 GMM techniques across the 500 Monte Carlo replications. For reasons of space we present only the results for the small sample size experiments ($T = 20$) and for the large sample sizes ($T = 200$). Results for the intervening cases ($T = 50$, 100 etc.) are available from the authors. In what follows, we take the bias in an estimated parameter to be an indicator of the relative *accuracy* of the given estimator in the specified circumstances, and the variance to be an indicator of the estimator's *reliability* (or sensitivity to "outliers" in the data).

TABLE 3

Beta dist. SAMPLE SIZE = 200	1st param. p		2nd param. q		average χ^2 p-value	
METHOD	bias	variance	bias	variance	(d.f. = 17)	
TRUE PARAM. $(\tau, \lambda) = (1, 3)$						
AHH	0.01305	0.00773	0.03151	0.09022	14.0	67%
HNW	0.01605	0.01079	0.03684	0.10727	14.2	
D–L	0.01697	0.01130	0.03940	0.11056	14.3	
DS	0.01730	0.01143	0.04027	0.11139	14.5	
SS(3)	0.03027	0.01874	0.07352	0.15800	14.7	
ML	0.04406	0.01424	0.12045	0.12527	15.0	59%
SIMPLE	–0.04658	0.04865	–0.17615	0.43037	16.0	
TRUE PARAM. $(p, q) = (1, 1)$						
AHH	–0.00261	0.00900	0.00081	0.00923	10.4	89%
HNW	0.01294	0.01027	0.01057	0.01067	11.8	
D–L	0.01331	0.01056	0.01480	0.01181	12.0	
DS	0.01341	0.01816	0.01491	0.01221	12.1	
SS(3)	0.01335	0.01886	0.01561	0.01243	12.2	
ML	0.01368	0.02017	0.01519	0.01121	12.6	71%
SIMPLE	0.65246	0.33157	0.64924	0.36377	14.2	
TRUE PARAM. $(p, q) = (3, 1)$						
AHH	0.05765	0.09784	0.01676	0.00908	13.0	74%
HNW	0.06025	0.12286	0.01872	0.01303	13.4	
D–L	0.06180	0.12636	0.01931	0.01356	13.6	
DS	–0.06201	0.12651	–0.01947	0.07660	14.0	
SS(3)	0.06253	0.12691	0.01958	0.01365	14.6	
ML	0.09112	0.17508	0.03061	0.02094	14.8	61%
SIMPLE	0.15116	0.13977	0.04759	0.01699	15.0	

(a) General results:

Both criteria, small sample bias and small sample efficiency, put our own GMM estimator (denoted AHH here) in first place for performance and the Hansen-Newey-West estimator (HNW) in second place. There are a total of 88 comparisons here[2], and there is just one case where our GMM estimator does not perform best [the maximum likelihood technique produces a marginally smaller variance for the second parameter estimate in the $N(2, 0.25)$ and $T = 200$ case]. Similarly there are just two cases

[2] 11 distributions (of 3 types) each with 2 parameters judged by 2 criteria in 2 sample size experiments.

Table 3 (continued)

SAMPLE SIZE = 20						
TRUE PARAM. $(p, q) = (1, 3)$					(d.f. = 4)	
AHH	0.15040	0.14359	0.60575	1.95128	4.5	34%
HNW	0.15425	0.18224	0.60940	2.25979	4.7	
D–L	0.40169	0.93666	1.23680	6.73138	5.3	
DS	0.15920	0.18723	0.62354	2.30046	4.8	
SS(3)	0.16143	0.18834	0.63016	2.30985	4.9	
ML	0.47709	0.46201	1.63302	5.00642	5.7	14%
SIMPLE	0.42583	1.47930	1.22201	9.34737	6.0	
TRUE PARAM. $(p, q) = (1, 1)$						
AHH	0.10146	0.14182	0.09315	0.17103	3.9	42%
HNW	0.13561	0.19563	0.12821	0.17347	4.2	
D–L	0.13698	0.19575	0.12991	0.18961	4.5	
DS	0.13584	0.19637	0.12863	0.18467	4.3	
SS(3)	0.14291	0.18025	0.13008	0.18806	4.8	
ML	0.24933	0.38327	0.24082	0.34987	5.2	17%
SIMPLE	1.28978	7.73841	1.15604	7.62929	7.2	
TRUE PARAM. $(p, q) = (3, 1)$						
AHH	0.47206	2.22174	0.18087	0.15325	5.3	26%
HNW	0.48975	2.38286	0.19356	0.18169	5.5	
D–L	1.38579	8.98914	1.16852	3.57749	6.0	
DS	0.50587	2.41933	0.19982	0.18717	5.8	
SS(3)	0.58747	2.99887	0.19781	0.18262	6.3	
ML	0.50126	2.41273	0.19806	0.18607	6.1	11%
SIMPLE	1.52819	5.69004	0.54230	0.48973	7.1	

where the Hansen-Newey-West estimator is not second best [the Deaton-Laroque method produces a lower bias for the second parameter in the $G(3,1)$ and $G(1,3)$, $T = 20$, cases]. After these two, the accuracy and reliability of the estimators deteriorate *rapidly*, especially with small sample sizes and in the Beta and Gamma distribution exercises.

A second general observation is that the estimators are differentiated more clearly in terms of reliability than in terms of accuracy: the variances of the parameter estimates show more variation over different techniques than do the small sample biases. This suggests that unbiasedness is a property which, within a given tolerance, is reached earlier than efficiency with expanding sample sizes. It further suggests that the asymptotic properties are an unreliable guide to the true parameter values at small sample sizes.

(b) The Normal Distribution Case (Table 1):

The most obvious feature of these results is that the methods are largely unbiased and efficient. An exception is the poor performance of the Deaton-Laroque estimator in small samples ($T = 20$). This poor performance is concentrated in the variances of these parameter estimates, which are often (but not always) 50 to 500 times larger than that of the other estimators – particularly for the second parameter. There are fewer problems with the bias of the estimates, although 7 out of the 10 bias results show signs opposite to the other estimators. Things look better in large samples. The Deaton-Laroque estimator generally produces better results than the methods of simulated moments or maximum likelihood and captures third place for large values of T. Consequently it appears that the Deaton-Laroque method requires much larger samples to achieve reasonable statistical properties. This sensitivity or unreliability in small sample sizes is also shared by the maximum likelihood estimates and also appears in the Beta and Gamma distribution results below, although the Deaton-Laroque estimator will not be singled out there since all the estimators (beyond the best two) do badly in those exercises.

A second feature is that both the bias and the variance of the estimates fall roughly by a factor of 10 with a 10-fold increase in the sample size, suggesting that estimates by any of the 7 methods converge on statistical consistency at the rate of $0(T^{-1})$. This feature does not seem to vary much among the different techniques. Thus, the performance ranking remains as described above: our GMM estimator dominates the Hansen-Newey-West estimator in every case, and the latter in turn dominates all others. Moreover the degree of dominance of our estimator over the Hansen-Newey-West estimator is usually larger than the dominance of the latter over the next best. Finally, both the bias and the variance of the estimates rise somewhat with larger values of σ^2 (the distribution's second parameter) and rather less so with μ (its first parameter), but these tendencies are weak compared to the results which follow in Tables 2 and 3.

The most awkward result in Table 1, therefore, is the poor performance of the maximum likelihood estimator. In this exercise, it produces independent estimates of μ and σ^2 and should be efficient at any sample size; its results should be at least as good as any of the others. One explanation why this is not so is that the differences observed in Table 1 may not be statistically significant but are simply the result of different numerical procedures. This possibility is examined in Section 5 below.

(c) The Gamma distribution case (Table 2):

The estimates in Table 2 show much greater bias and inefficiency – particularly in small samples, where the estimates of at least one of the two parameters are really very poor. It seems that consistency here requires a substantially larger sample size than for problems involving normally distributed variables.

Having said that, our GMM estimator still dominates Hansen-Newey-West, but by a smaller margin than the latter dominates the next best. In that sense, the best two both pull ahead of the pack in small samples. This implies that there is an increasing *relative* (but not absolute) reliability as the quality of the estimates starts to fall. In any event, it now matters more which estimator one chooses. Moreover, the choice between estimators is wider in that, even in larger samples, the biases and variances of the parameter estimates may be 8–10 times larger if the "wrong" estimator is used. Once again, both the Deaton-Laroque and the maximum likelihood estimators appear to be more unreliable and inaccurate than the others in small samples. On the other hand, the biases and variances from the two best techniques fall by factors of 10 or more when the sample size is increased from 20 to 200, suggesting that the estimators are still converging on consistency a little faster than $0(T^{-1})$.

Finally, both the bias and variances tend to increase with the size of the underlying parameter but, interestingly, not with the size of the other parameter value.

(d)The Beta distribution case (Table 3):

Like in the Gamma distribution results, the estimates in Table 3 show relatively large biases and variances in small samples. Consistency therefore requires fairly large samples, though not as large as the Gamma distribution estimates.

As before, our GMM estimator dominates all others, and the Hansen-Newey-West estimator comes second, for accuracy, reliability, and mean square errors. The degree of dominance is reasonably large again, which is consistent with the proposition that these two estimators pull ahead of the pack as the quality of the estimates starts to fall. The simple method of moments and the Deaton-Laroque estimators continue to perform badly in small samples, and the two best methods show biases and variances falling by factors of 10 or more as the sample increases from 20 to 200. So once again, consistency appears to be achieved at a rate of more than $0(T^{-1})$.

5. TESTING THE SIGNIFICANCE OF THE OBSERVED SMALL SAMPLE BIASES

The most striking feature of these results is the relatively poor performance of the maximum likelihood estimators. Numerically, the maximum likelihood estimators produce the worst or second worst bias results in 19 out of the 20 Normal distribution tests (Table 1), 6 out of the 12 Gamma distribution tests (Table 2), and 9 out of 12 Beta distribution tests (Table 3). Similarly they show the largest or second largest variances in 18 out of 20 tests in Table 1, in 2 of the 12 tests in Table 2, and in 6 out of 12 tests in Table 3. So, for accuracy and reliability, maximum likelihood performs relatively badly compared to the leading GMM estimators.

These results are remarkable because we know that, in the case of normally distributed variables at least, maximum likelihood estimates are independently distributed and efficient in the sense of actually reaching the Cramer-Rao lower bound (Mood, Graybill and Boes, 1974, chapter 7). Theoretically, they cannot be beaten.

For the Gamma and Beta distribution, the theory is not so clear. First, maximum likelihood estimates are now no longer independent of one another. Estimators that pay little attention to the higher order moments of the distribution being fitted may be able to secure lower biases (or variances) in their parameter estimates at the implicit cost of higher biases in those higher order moments which are unpenalised. Such trade-offs are not available for an estimator that tries to fit the entire likelihood function.

Second, the unbiasedness and efficiency properties of maximum likelihood estimation are now only asymptotic, and our sample sizes of 20 to 200 may be too small to capture those properties. Thus maximum likelihood estimates may have produced worse results than some of the GMM estimators because the maximum likelihood estimators' sampling distributions converge more slowly (in both mean and variance) to their limiting distribution.

These arguments can explain why the maximum likelihood estimates are worse than some of the GMM estimates in the Gamma and Beta tests, but not why they are worse in the Normal distribution tests. In this latter case, there are only two possible explanations: the maximum likelihood biases (variances) in Table 1 are not statistically significant, and/or they are the result of numerical instabilities in the algorithm used to compute them. The implications of these two explanations are quite different, however. If the biases (variances) are not significantly different from zero (or each other), then there are not problems with the estimating techniques, and, statistically it does not matter which is chosen. But if they are significant, it would be worthwhile to examine the numerical properties of different maximum likelihood algorithms to eliminate (as far as possible) any problems of numerical instability.

(a) The results from the Normal Distribution tests

The distribution of the maximum likelihood estimates of the parameters of a normal distribution that actually generate the observations is straightforward. For a sample size of T, and for N replicated samples, the first parameter would be $\bar{x}_i = T^{-1}\sum_j x_{ij}$ for the ith sample, and the bias would be computed from the average of N replications, $\bar{x} = N^{-1}\sum \bar{x}_i = (NT)^{-1}\sum\sum x_{ij}$. If the underlying observations x_{ij} are drawn from a Normal(μ, σ^2) distribution, then $\bar{x} \sim N(\mu, \sigma^2/NT)$ *exactly*, and our recorded bias is distributed as $N(0, \sigma^2/NT)$ in Table 1. Similarly our recorded estimate of the second parameter (the distributions variance) is the average of N replications; i.e. $s^2 = N^{-1}\Sigma s_i^2$, where each $(T-1)s_i^2/\sigma^2$ is distributed $\chi^2_{(T-1)}$ and has variance $2(T-1)$. Hence s_i^2 has variance $2\sigma^4/(T-1)$ and $s^2 \overset{a}{\sim} N(\sigma^2, 2\sigma^4/N(T-1))$, so bias in the latter is distributed asymptotically as $N(0, 2\sigma^4/N(T-1))$.

For the results in Table 1, we have $T = 20$ or 200, $N = 500$, and five different generating distributions. The bias in the first parameter estimate (Table 1, column 1) will therefore be significantly different from zero at a 5% level if it lies outside the interval $\pm 1.96\sigma_{\bar{x}}$ where $\sigma_{\bar{x}} = \sigma/\sqrt{NT}$. Similarly the biases in the second parameter estimate will be significant at a 5% level if they lie outside the interval $\pm 1.96\sigma_s$, where $\sigma_s = \sigma^2\sqrt{2/N(T-1)}$. Table 4 summarises $\sigma_{\bar{x}}$ and σ_s for the five different distributions represented in Table 1.

Evidently the maximum likelihood estimates of the mean (μ) show significant biases at a 5% level only in the $N(0, \frac{1}{4})$ case with $T = 200$. The remaining 9

TABLE 4

	$\sigma_{\bar{x}}$			σ_s		
	$\sigma^2 = 0.25$	$\sigma^2 = 1$	$\sigma^2 = 2$	$\sigma^2 = 0.25$	$\sigma^2 = 1$	$\sigma^2 = 2$
$T = 20$	.005	.01	.014	.004	.015	.029
$T = 200$	.0016	.0032	.0045	.001	.005	.009

maximum likelihood estimators show no significant biases. These tests are exact. The tests of bias in the estimates of the variance parameter are asymptotic with respect to a "sample" size of 500, but show a greater number of significant biases. In fact, significant biases appear in all 11 second parameter estimates.

Hence maximum likelihood estimation has produced some significant biases, more when estimating of the second parameter than the first. Both the presence of significant biases in half the cases and the fact that these biases tend to appear in both the variance parameter and the larger samples for the estimated mean suggest that numerical instability is at least part of the reason for the poor maximum likelihood performance. Indeed the estimates for $\sigma_{\bar{x}}^2$ and σ_s^2 from Table 4 are smaller than the variances of the estimates actually recorded in columns 2 and 4 of Table 1. The computed distributions of our estimates are therefore very much wider than the Cramer-Rao lower bound would imply, which is symptomatic of numerical instability.

However, that is not the real issue. The crucial question is, are these biases actually larger (in a statistical sense) than those coming from the GMM estimators? The difficulty here is that the exact distributions of the parameter estimates obtained from the GMM techniques are not analytically tractable, since their estimating equations do not admit a closed form solution. Thus, we cannot obtain an exact variance of the parameter estimates $\bar{x}_i$ and s_i^2 to which we could apply the central limit theorem to derive tests of the biases in $\bar{x}$ or s^2. However we can use the maximum likelihood values already obtained to estimate those variances. That gives the test results in Table 7.

Thus, whereas it is possible to argue that maximum likelihood estimators do not provide any significant biases in the first parameter, and that the actual biases observed are the result of numerical instabilities in the algorithm used to maximise the likelihood function, the same cannot be said for the GMM estimators. With our tests, there is a much higher incidence of significant bias in both parameters and both sample sizes. The chief offenders are the Deaton-Laroque (DL) and the Method of Simulated Moments (DS) estimators. At the other end of the scale, our own GMM estimator and the HNW estimator produced no significant biases in the first parameter estimate and fewer in the second. Hence there are significant differences in accuracy and reliability between the AHH and HNW estimators on the one hand, and the remaining GMM estimators on the other.

(b) The Gamma Distribution Tests

Here formal testing for bias is difficult since the exact distribution of estimates, even for maximum likelihood, of the two parameters is unknown. Thus, we are unable to determine the variances of those estimators to which the Central Limit Theorem might otherwise apply. Further we cannot substitute the estimated variances (maximum likelihood or otherwise) obtained in Table 2, for biased parameter estimates entail biased variance estimates (there being no independence property now). Any formal justification for this approach has thus disappeared.

But even if conventional asymptotic tests are not possible, a *conditional* test can be used that is a sufficient (but not necessary) condition for detecting significant biases. The maximum likelihood estimates of a Gamma distribution are obtained by solving

$$\left.\begin{aligned} &\tau = \bar{x}\lambda \\ &and\ \lambda = e^{\psi(\tau)} \Big/ \left(\prod_{i=1}^{T} x_i\right)^{1/T} \end{aligned}\right\} \tag{4}$$

simultaneously for τ and λ, where $\bar{x} = T^{-1} \sum_{i=1}^{T} x_i$, and the x_i are the random drawings in a sample of size T. The function $\psi(\tau)$ is the Digamma function: $\psi(\tau) = \text{dlog}\Gamma(\tau)/d\tau$ where $\Gamma(\tau) = (\tau - 1)!$. Note that $\psi(\tau) > 0$ is monotonically increasing in τ for $\tau \geq 1$.

However, for testing purposes, we can form conditional estimates of τ and λ by inserting the true values of λ and τ from the underlying distribution on the right of (1). Call these conditional estimates τ^* and λ^*, and let the actual estimates obtained by solving (1) be $\hat{\tau}$ and $\hat{\lambda}$. Then, with positively biased estimates, the probability of any particular positive bias in $\hat{\tau}$ or $\hat{\lambda}$ is *less* than the probability of the same bias in τ^* or λ^* under the null of unbiasedness. Hence a sufficient condition for the biases in Table 2 to be significant (at the 5% level) is that they should be significant for τ^* or λ^*. Indeed (4) implies that $\tau* \overset{a}{\sim} (\tau, \lambda^2\sigma^2/NT)$. Using the fact that $\sigma^2 = \tau/\lambda^2$ for each of the three Gamma distributions estimated, we find the maximum likelihood estimates of τ to be significantly biased[3] for both the small and large samples. The consistency and asymptotic efficiency of GMM estimators allow us to extend these asymptotic tests to the other estimators of τ in Table 2. Once again, all estimates show significant biases.

Hence we conclude that, in the case of the Gamma distribution tests, all estimators show significant biases that do not vanish with larger sample sizes. The sampling distributions evidently converge slowly on their asymptotic distributions, in terms

---

3

Note: $\lambda^2\sigma^2/NT$ for	$G(3,1)$	$G(1,3)$	$G(1,1)$
$T = 20$	.017	.01	.01
$T = 200$	.0055	.0032	.0032

TABLE 5
Biases in the estimated means and variances of gamma distributed variables from Table 2 ($\mu = \hat{\tau}/\hat{\lambda}$; $\sigma^2 = \hat{\tau}/\hat{\lambda}^2$).

True parameters and distribution	Estimator	$T = 20$ Bias in mean	$T = 20$ Bias in variance	$T = 200$ Bias in mean	$T = 200$ Bias in variance
(3,1)	AHH	.008	–.452	.0113	–.02
$\mu = 3$	HNW	.007	–.529	.0115	–.034
$\sigma^2 = 3$	DL	.145	–.379	–.2015	.225
	DS	–.212	–.958	–.0106	–.188
	SS(3)	.124	–.987	.040	–.131
	ML	.029	–1.006	.0142	–.147
	Simple	–.029	–1.330	–.016	–.331
(1,3)	AHH	–.018	–.0234	.0006	–.0012
$\mu = 1/3$	HNW	–.018	–.0332	.0007	–.0027
$\sigma^2 = .11$	DL	–.001	–.0286	.0005	–.0049
	DS	–.016	–.0386	.0050	.0090
	SS(3)	–.013	–.0430	.0008	–.0085
	ML	–.013	–.0560	.0080	–.0179
	Simple	–.021	–.0596	.0018	–.0090
(1,1)	AHH	–.055	–.215	.0017	–.011
$\mu = 1$	HNW	–.055	–.303	.0012	–.025
$\sigma^2 = 1$	DL	–.015	–.410	.0141	.081
	DS	–.040	–.460	.0103	–.059
	SS(3)	–.016	–.468	.0039	–.082
	ML	–.004	–.421	–.0260	–.167
	Simple	–.08	–.583	.0097	–.098

both of unbiasedness and having larger variances than in the limit (compare values for $\lambda^2\sigma^2/NT$ with column 2 of Table 2). It is clear that both our preferred GMM estimators (AHH first, and then HNW) are more accurate and more reliable (having smaller biases and lower variances) than their rivals – including maximum likelihood. But, this does not cause them to be unbiased or near-minimum variance. In fact these results are purely relative: while our own GMM estimator is preferable to the others, it is not necessarily good.

And this is as far as we can go. Conditional tests on λ itself are not possible since the variance for the distribution of the inverse geometric mean, $(\Pi x_i)^{-1/T}$, is not known and the central limit theorem cannot be applied. Beyond this, we can only look at the biases in the estimated means ($= \hat{\tau}/\hat{\lambda}$) and variances ($= \hat{\tau}/\hat{\lambda}^2$)

numerically. These figures are given in Table 5, but formal tests are not possible since both are derived from ratios of nonindependently distributed random variables. It is clear from Table 5 that the biases in the mean are systematically smaller than those in the variance, and they vary less across estimators than do those for the variance estimates.[4]

These results illustrate an important point. General statements indicating that a particular estimator is more accurate, or converges faster to its asymptotic distribution, can be extremely misleading. In this exercise the means have been well estimated in all cases. The variances are less well estimated – but their fit is still good compared to many of the estimates of the τ, λ parameters. And such results are easily obtained, since even significant biases in τ and λ of the same sign will offset each other to produce means or variances with relatively little bias. That is, the quality of the results obtained from estimating particular characteristics may be quite different from those obtained from fitting the distribution as a whole. Hence it matters whether the real objective is to fit particular parameters or the distribution as a whole.

(c) The Beta Distribution Tests

Here not even conditional tests are available to determine the significance of the biases in the maximum likelihood estimates of Table 3. These estimates arise from solving

$$\left.\begin{aligned} T[\psi(p+q) - \psi(p)] + \sum \log x_i = 0 \\ and\ T[\psi(p+q) - \psi(q)] + \sum \log(1 - x_i) = 0 \end{aligned}\right\} \tag{5}$$

simultaneously for p and q, a process that does not yield a tractable closed-form solution. At best one can inspect the numerical biases in Table 3 or the equivalent bias results in Table 6. But, just as these, Table 6 shows how easily numerically "significant" biases in the parameter estimates can offset one another to give apparently unbiased mean and variance estimates. Both are estimated with much smaller numerical biases than are p and q themselves. There is no clear tendency here for the variance to be more biased than the mean, and both biases show a stronger tendency to diminish with increasing T. Nor is there any apparent ranking of biases across estimators. Yet the general message is the same: it matters for estimation whether one focuses on particular characteristics of the distribution or its entirety.

6. RESULTS: FITTING THE ENTIRE DISTRIBUTION

To test the goodness of fit of the entire distribution implied by each replication underlying the results in Tables 1 to 3, we have used the traditional χ^2 test: the likelihood ratio goodness-of-fit tests (Kendall and Stewart (1974)). The mean χ^2

---

[4] Our own GMM estimator generally does better than the other estimators in Table 5. On the other hand the bias in the variance estimates converges to zero with increasing T, but there is little convergence of the biases in the means.

TABLE 6

Biases in the estimated means and variances of beta distributed variables from Table 3.

$$\left(\mu = \frac{p}{p+q},\ \sigma^2 = \frac{pq}{(p+q)^2\,(p+q+1)}\right)$$

True parameters and distribution	Estimator	$T = 20$ Bias in mean	$T = 20$ Bias in variance	$T = 200$ Bias in mean	$T = 200$ Bias in variance
(1,3)	AHH	.0081	–.0056	.0005	.0003
$\mu = 1/4$	HNW	.0077	–.0056	.0007	–.0048
$\sigma^2 = .0375$	DL	.0014	–.0094	.0007	–.0003
	DS	.0076	–.0057	.0007	–.0004
	SS(3)	.0076	–.0058	.0011	–.0006
	ML	.0083	–.0117	.0007	–.0011
	Simple	.0025	–.0091	.0024	.0002
(1,1)	AHH	–.0377	–.0109	–.0009	.0001
$\mu = 1/2$	HNW	–.0583	–.0156	.0006	–.0003
$\sigma^2 = .083$	DL	.0016	–.0065	–.0004	–.0004
	DS	.0016	–.0064	–.0004	–.0004
	SS(3)	.0028	–.0066	–.0006	–.0005
	ML	.0017	–.0114	–.0004	–.0005
	Simple	.0150	–.0371	.0005	–.0249
(3,1)	AHH	.0038	–.004	.0005	–.0006
$\mu = 3/4$	HNW	.0049	–.004	.0003	–.0006
$\sigma^2 = .0375$	DL	.0396	–.003	.0002	–.0006
	DS	.0050	–.004	–.0002	.0007
	SS(3)	.0003	–.005	.0002	–.0006
	ML	.0049	–.004	–.0001	–.0009
	Simple	.0041	–.0107	.0005	–.0015

statistics, for each estimation technique under review, are given in Tables 1 to 3. The conventional goodness of fit test would accept the null hypothesis that the observations fitted by the named technique conformed to a normal, gamma or beta distribution, respectively, if the associated χ^2 test statistics were less than the critical values of 27.6 (for a 5% significance level and $T = 200$). For $T = 20$, the critical value is 9.5. Every estimator therefore passes this test easily, even in the smaller samples, and the null hypothesis is correctly accepted.

It is clear, however, that these tests are considerably more powerful in the larger

TABLE 7

Test the significance of the estimates' bias for Normal (μ, σ^2)

	$N(0,1)$		$N(0,2)$		$N(0,1/4)$		$N(2,2)$		$N(2,1/4)$	
$T = 200$	$\hat{\mu}$	$\hat{\sigma}^2$	$\hat{\mu}$	$\hat{\sigma}^2$	$\hat{\mu}$	$\hat{\sigma}^2$	$\hat{\mu}$	$\hat{\sigma}^2$	$\hat{\mu}$	$\hat{\sigma}^2$
AHH	–	–	–	*	–	–	–	*	–	*
HNW	–	*	–	*	–	*	–	*	–	*
DL	–	*	–	*	*	*	*	*	*	*
DS	–	*	–	*	*	*	–	*	–	*
SIMPLE/ SS(3)/ML	–	*	–	*	*	*	–	*	–	*
$T = 20$										
AHH	–	*	–	*	–	*	–	*	–	*
HNW	–	*	–	*	–	*	–	*	–	*
DL	–	*	*	*	*	*	–	*	*	*
DS	–	*	–	*	–	*	–	*	–	*
SIMPLE/ SS(3)/ML	–	*	–	*	–	*	–	*	–	*

Note: * indicates a significant bias at the 5% level.

samples. Indeed, although we have not specified a particular alternative hypothesis, the estimator producing the lowest calculated χ^2 statistic minimises the probability of making a type II error for any given alternative hypothesis, whatever it may be. For larger samples the observed significance level, or p-value, corresponding to the calculated χ^2 statistic ranges from 72% to 89% in the Normal distribution exercises for the best of our estimators. This range is from 67% to 89% in the Beta and Gamma distribution cases, placing the conventional 5% or 10% critical values. For the small samples, the p-values are lower: 34% to 54% in Table 1, 21% to 43% in Table 2, and 34% to 54% in Table 3.

These results also confirm the performance ranking established in the previous section. In all 11 experiments, and for both sample sizes, our own GMM estimator produced a distribution that matched the true distribution better than any of the distributions fitted by the other estimators. The Hansen-Newey-West estimator came in second place again, followed by Deaton-Laroque, the method of simulated moments, and the maximum likelihood estimator. Moreover, the difference in the χ^2 test statistics between the best GMM estimator and the maximum likelihood estimator indicates an improvement of between 8% and 40% in the p-value or confidence level for accepting the null hypothesis that the estimated distribution successfully fits the specified distribution in large samples, and an improvement of between 14% and 32% for the smaller samples. This is a healthy finite sample improvement over traditional estimation methods.

7. CONCLUSIONS

Basically, our concerns about the poor small sample properties of GMM estimators have been born out. While we have observed a fairly rapid rate of convergence towards consistency and asymptotic efficiency, there is still evidence of statistically significant biases and large variances, even in the larger samples.

Just how bad the small sample properties actually are depends on the particular estimation technique chosen. It matters which GMM estimator is used and which numerical implementation of the maximum likelihood estimator is applied. In these exercises there is a clear ranking: our own GMM estimator performs best, followed by the Hansen-Newey-West estimator, and then the Method of Simulated Moments. The Deaton-Laroque estimator shows a great deal of variability in small samples, but is a relatively good performance in larger samples.

Moreover, it appears that the differences between the performance of these estimators widen as we depart from the classical assumptions of large samples and normally distributed variables. We find the results are sensitive to the sample size, the form of fitting criterion, non-normality in the underlying distribution, and the size of the parameter being estimated. We also find that most estimators are worse in regard of efficiency than unbiasedness. Nevertheless, the GMM estimators all fairly good for fitting probability distributions in their entirety, even in relatively small samples.

APPENDIX

THEORETICAL MOMENTS UNDER DIFFERENT DISTRIBUTIONS

(1) Normal distribution

$$\text{p.d.f: } f(X,\beta) = \frac{1}{\sqrt{2\pi\sigma^2}} \exp\left(-\frac{(x-\mu)^2}{2\sigma^2}\right)$$

The $\mu_1 = \mu$, $\mu_2 = \sigma^2$, $\mu_3 = 0$, $\mu_4 = 3\sigma^4$.

(2) Gamma distribution

$$\text{p.d.f: } f(X,\beta) = \frac{\lambda^\tau}{\Gamma(\tau)} X^{\tau-1} e^{-\lambda X} \quad X, \tau, \lambda > 0$$

where

$$\Gamma(\tau) = \int_0^\infty s^{\tau-1} e^{-s}\, ds$$

Then $\mu_1 = \tau/\lambda$, $\mu_2 = \tau/\lambda^2$, $\mu_3 = 2\tau/\lambda^3$, $\mu_4 = 3\tau(\tau+2)/\lambda^4$.

(3) Beta distribution

$$\text{p.d.f: } f(X,\beta) = \frac{\Gamma(p+q)}{\Gamma(p)\,\Gamma(q)}\, X^{p-1}(1-X)^{q}\,,\ \ 0 < X < 1\,,\ \ p,q > 0.$$

Then

$$\mu_1 = \frac{p}{p+q} \quad \mu_2 = \frac{pq}{(p+q+1)\,(p+q)^2}$$

$$\mu_3 = \frac{2pq(q-p)}{(p+q+2)\,(p+q+1)\,(p+q)^3}$$

$$\mu_4 = \frac{3pq(p^2q + 2p^2 - 2pq + pq^2 + 2q^2)}{(p+q)^4\,(p+q+1)\,(p+q+2)\,(p+q+3)\,(p+q+4)}$$

ACKNOWLEDGEMENTS

We are grateful to Dave Belsley, Gregor Smith, Jim Powell, Robin Lumsdaine and participants of the Econometrics Seminar at Princeton for their comments.

REFERENCES

Deaton, A.S. and Laroque, G. (1992) On the behaviour of commodity prices, *Review of Economic Studies*, 59, 1–24.

Duffie, D. and Singleton K.J. (1989) Simulated Moments Estimation of Markov Models of Asset Prices, Stanford University Discussion Paper, Stanford, CA.

Gregory, A. and G. Smith (1990) "Calibration as Estimation" *Econometric Reviews*, 9, pp.57–89.

Hansen, L.P. (1982) Large sample properties of generalised Method of Moments Estimators, *Econometrica*, Vol. 50, pp 1029–1054.

Hughes Hallett, A.J. (1992) Stabilising earnings in a volatile market, paper presented in the Royal Economics Society Conference, London (April).

Kendall, M.G. and Stewart, A. (1973) *The Advanced Theory of Statistics*, Vol. 2, Third Edition, Griffen & Co., London.

Mood, A. F. Graybill and D. Boes (1974) *Introduction to the Theory of Statistics*, McGraw-Hill, New York.

Newey, W.K. and West K.D. (1987) A Simple, positive semi-definite, heteroscedasticity and autocorrelation consistent covariance matrix, *Econometrica*, 55, pp 703–708.

Smith, G. and Spencer M. (1991) Estimation and testing in models of exchange rate target zones and process switching, in P. Krugman and M. Miller (eds), *Exchange rate targets and currency bands*, Cambridge University Press, Cambridge and New York.

Tauchen, G. (1986) Statistical Properties of Generalised Method of Moments Estimators of Structural Parameters Obtained from Financial Market Data, *Journal of Business and Economic Statistics*, 4, pp.397–425.

ALBERT J. REED AND CHARLES HALLAHAN

A Bootstrap Estimator for Dynamic Optimization Models

ABSTRACT. We propose a technique for computing parameter estimates of dynamic and stochastic programming problems for which boundary conditions must be imposed. We demonstrate the feasibility of the technique by computing and interpreting the estimates of a dynamic food price margin model using secondary economic time series data.

1. INTRODUCTION

Several solutions to infinite time-horizon, multivariate stochastic and dynamic programming problems have recently been proposed (Baxter et al. 1990; Christiano, 1990; Coleman, 1990; den Haan and Marcet, 1990; Gagnon, 1990; Labadie, 1990; McGratten, 1990; Tauchen, 1990; Taylor and Uhlig, 1990). However, few studies suggest ways to make correct inferences on parameter estimates in such problems. An exception has been the recent work of Miranda and Glauber (1991). The complex restrictions that the coefficients of such solutions must obey can inhibit statistical inference. Simplifying the restrictions requires simplifying the model structure, and inferences on a simplified model may only be of limited use. Alternatively, inferences can be made from the first-order conditions of the problem. However, this strategy forces the analyst to ensure that the parameter estimates satisfy the problem's boundary conditions. In Miranda and Glauber, (1991) boundary conditions are inherited through price band policies. Our study applies to the problem of estimating the parameters of a dynamic problem in which no inherent boundary conditions exist, but for which economic theory requires certain restrictions to be satisfied if the model is to be useful in explaining behavior.

We illustrate our method with a stochastic regulator problem. This optimization framework embodies linear-quadratic models (Sargent, 1987a) and provides the economic arguments that underly some vector autoregression models. It also can be used to approximate dynamic optimization problems without closed form solutions (McGratten). Our study suggests how one could make (approximately) correct statistical inferences on a model whose parameters satisfy fixed point or boundary conditions.

Gallant and Golub (1984) illustrate how one could impose inequality restrictions on a static optimization problem. Using their methodology, one could impose restrictions on the eigenvalues of the matrices of the stochastic regulator, thereby achieving the required boundary condition. However, such a strategy places more restrictions on the parameter estimates than the boundary condition.

Our procedure also can be used to estimate the parameters of static optimization

D. A. Belsley (ed.), Computational Techniques for Econometrics and Economic Analysis, 45–63.

problems, but we apply it here to dynamic and stochastic problems. The stochastic regulator encompasses a wide range of dynamic and stochastic models, and dynamic and stochastic models provide a rich interpretation of economic data. These models readily differentiate among the response of an endogenous variable to an actual change, to a perfectly expected change, and to an unexpected change in an exogenous variable. Furthermore the problem addresses the Lucas critique by recognizing that such responses are not invariant to systematic changes in policy.

After discussing the stochastic regulator problem in Section 2, the bootstrap estimator is presented in Section 3. Section 4 provides an example of interest to agricultural economists, and Section 5 summarizes the paper.

2. THE STOCHASTIC OPTIMAL REGULATOR PROBLEM

Here we review the setup of the stochastic regulator, its solution, and the conditions that deliver the solution. A more thorough treatment can be found in Sargent (1987b, Chapter 1). An understanding of the stochastic regulator is crucial to understanding the estimation procedure.

Consider a general dynamic and stochastic optimization problem defined by a vector of state variables $x = [x_1' : x_2']'$ and a vector of control variables u. The problem is to find the control sequence $\{u_t\}$ satisfying

$$V(x_0) = \max_{\{u_t\}} \mathcal{E}_0 \sum_{t=0}^{\infty} \beta^t \pi(x_t, u_t)$$

subject to x_{-1}, and the equations of motion

$$x_{1t+1} = g_1(x_t, u_t) ,$$

$$x_{2t+1} = g_2(x_{2t}, \varepsilon_{t+1}) ,$$

and the probability distribution

$$\text{Prob}(\varepsilon_t < e) = G(e) .$$

Here the vector x_1 is termed the 'endogenous' state variable, x_2 the 'exogenous' state variable, and ε_t is a serially uncorrelated error term satisfying $\mathcal{E}(\varepsilon_t | x_t, x_{t-1}, \ldots; \varepsilon_{t-1}, \varepsilon_{t-2}, \ldots) = 0$. $V(x_0)$ is the value or objective function in period 0, $\pi_t(x_t, u_t)$ is the return function in period t, and β is the discount factor. $\mathcal{E}_t(y)$ denotes the mathematical expectation of the random variable y conditioned on the state variable in time t, and taken with respect to G.

Two features characterize the above infinite time horizon problem. First, x_{2t+1} does not depend on x_{1t} or u_t. Thus, x_{1t} does not Granger cause x_{2t}. Second, the problem is recursive. The selection of u in the current period affects current and future period returns and future period x_1 without affecting past-period returns and past x_1. This recursivity enables the analyst to re-cast the above infinite-time-horizon problem as a two-period problem that can be solved sequentially.

Specifically, the recursive problem can be written as

$$V(x_t) = \max_u \{\pi(x_t, u_t) + \beta\mathcal{E}[V(x_{t+1})]|x_t\}$$

$$= \max_u \{\pi(x_t, u_t) + \beta\mathcal{E}[V(g(x_t, u_t, \varepsilon_{t+1}))]|x_t\}$$

subject to,

$$x_{1t+1} = g_1(x_t, u_t)$$

$$x_{2t+1} = g_2(x_{2t}, \varepsilon_{t+1})$$

where,

$$\mathcal{E}\, V(g(x_t, u_t, \varepsilon_{t+1})|x_t) = \int V(g(x_t, u_t, \varepsilon_{t+1}))\, dG(\varepsilon)\ ,$$

and $g = [g_1' : g_2']'$. The necessary conditions for a solution are:

$$\frac{\partial\pi(x_t, u_t)}{\partial u_t} + \beta\mathcal{E}\Big[\left(\frac{\partial g_1}{\partial u_t}, \frac{\partial g_2}{\partial u_t}\right) V'(x_{t+1})|x_t\Big] = 0\,.$$

For interior solutions, the value function satisfies

$$V'(x_{t+1}) = \begin{pmatrix} \dfrac{\partial\pi}{\partial x_{1t+1}} \\ \dfrac{\partial\pi}{\partial x_{2t+1}} \end{pmatrix} + \beta\mathcal{E} \begin{bmatrix} \dfrac{\partial g_1}{\partial x_1} & \dfrac{\partial g_2}{\partial x_1} \\ \dfrac{\partial g_1}{\partial x_2} & \dfrac{\partial g_2}{\partial x_2} \end{bmatrix} \begin{pmatrix} \dfrac{\partial V}{\partial x_{1t+2}} \\ \dfrac{\partial V}{\partial x_{2t+2}} \end{pmatrix}.$$

If

$$\frac{\partial g_1}{\partial x_1} = 0$$

and x_{1t} does not Granger cause x_{2t}, (i.e., $\partial g_2/\partial x_1 = 0$), then because $\partial g_2/\partial u_t = 0$, the necessary conditions reduce to

$$\frac{\partial\pi(x_t, u_t)}{\partial u_t} + \beta\mathcal{E}\left\{\left(\frac{\partial g_1}{\partial u_t}\frac{\partial\pi}{\partial x_{1,t+1}}\right)|x_t\right\} = 0\,.$$

The above conditions are termed Euler equations and have a convenient structure. The parameters of the Euler equations contain only the parameters of $\partial g_1/\partial u_t$, β, and the parameters of the return function. Unlike the first-order conditions for more general problems, the Euler equations do not contain $V'(x_{t+1})$, which complicates estimation efforts because changes systematically over an iterative solution procedure and presumably over a data sample. For this reason the proposed estimation procedure applies to dynamic problems in which the state and control variables can written so that $\partial g_1/\partial x_1 = 0$.

The Euler equations are unobservable because of the expectations operator. If e_t is a forecast error and x_{t-j} $(j = 0, 1, \ldots)$ are elements of an information set, the Rational Expectations Hypothesis (REH) states $\mathcal{E}(e_t|x_t, x_{t-1}, \ldots) = 0$. If the parameters of the Euler equations can be expressed in terms of the parameter vector θ, the forecast error is

$$\frac{\partial\pi(x_t,u_t)}{\partial u_t}+\beta\left[\frac{\partial g(x_t,u_t,\varepsilon_{t+1})}{\partial u_t}\,\frac{\partial\pi(x_{t+1},u_{t+1})}{\partial x_{1t+1}}\right]=e_t(\theta)\;.$$

The above relationships are referred to as the sample Euler equations. Notice that $\mathcal{E}(e_t|x_t,x_{t-1},\ldots)=0$ implies $\mathcal{E}e_t x_{t-j}=0$ $(j=0,1,\ldots)$. Hence, if one defines a vector of instruments z_t that consist of elements x_{t-j} $(j=0,1,\ldots)$, then

$$\mathcal{E}\left\{\left(\frac{\partial\pi(x_t,u_t)}{\partial u_t}+\beta\left[\frac{\partial g(x_t,u_t,\varepsilon_{t+1})}{\partial u_t}\,\frac{\partial\pi(x_{t+1},u_{t+1})}{\partial x_{1t+1}}\right]\right)\otimes z_t\right\}=\mathcal{E}e_t\otimes z_t=0\,.$$

where '$\otimes$' denotes the Kronecker product.

The above expression is the orthogonality condition exploited when computing Generalized Method of Moments (GMM) estimates of the parameters of the Euler equations. In particular, for n observations, the GMM estimate is (Gallant, 1987)

$$d=\text{argmin}_\theta\ S(\theta,\hat{V})\,,$$

where

$$S(\theta,\hat{V})=[nm_n(\theta,x)]'\,\hat{V}^{-1}[nm_n(\theta,x)]$$

and

$$m_n(\theta,x)=n^{-1}\sum_{t=1}^{\infty}e_t(\theta)\otimes z_t\;.$$

To obtain a closed-form solution to recursive dynamic and stochastic optimization problems, one must compromise on the functional form. The Stochastic Optimal Linear Regulator specifies a quadratic objective function and linear constraints. This class of models takes the form

$$V(x_0)=\max_{\{u_t\}}\ \mathcal{E}_0\sum_{t=0}^{\infty}\beta^t\left\{[x_t',u_t']\begin{bmatrix} r & w\\ w' & q\end{bmatrix}\begin{bmatrix} x_t\\ u_t\end{bmatrix}\right\}$$

subject to

$$x_{1t+1}=a_{11}x_{1t}+a_{12}x_{2t}+b_1u_t\;,$$

$$x_{2t+1}=a_{21}x_{1t}+a_{22}x_{2t}+b_2u_t+\varepsilon_{t+1}\;,$$

where

$$x_t=\begin{bmatrix} x_{1t}\\ x_{2t}\end{bmatrix}\;,\quad r=\begin{bmatrix} r_{11} & r_{21}\\ r_{21} & 0\end{bmatrix}\;.$$

The infinite time horizon problem is re-cast as a two period problem comprising Bellman's equation

$$V(x_0)=\max_{u_t}\left\{(x_t',u_t')\begin{bmatrix} r & w\\ w' & q\end{bmatrix}\begin{pmatrix} x_t\\ u_t\end{pmatrix}+\beta\mathcal{E}\,V(x_{t+1})|x_t\right\}$$

and the constraints

$$x_{1t+1} = a_{11}x_{1t} + a_{12}x_{2t} + b_1 u_t ,$$

$$x_{2t+1} = a_{21}x_{1t} + a_{22}x_{2t} + b_2 u_t + \varepsilon_{t+1} .$$

Assuming x_{1t} does not Granger cause x_{2t}, we have $a_{21} = 0$ and $b_2 = 0$, and the problem is recursive. If, in addition $a_{11} = 0$, the Euler equations

$$w'x_t + qu_t + \beta b_1' \mathcal{E}[r_{11}x_{1t+1} + r_{21}x_{2t+1}|x_t] = 0$$

serve as a set of necessary conditions. Notice the Euler equations only contain the parameters b_1 and the parameters of the objective function.

Now, define $b = [b_1' : b_2']'$, the matrix $\mathbf{a}$ with elements a_{ij}, and make the transformations

$$v_t = q^{-1}w'x_t + u_t$$

$$R = r - wq^{-1}w'$$

$$Q = q$$

$$A = a - bq^{-1}w'$$

$$B = b .$$

This permits the problem to be re-stated as

$$V(x_t) = \max_{v_t} \{x_t'Rx_t + v_t'Qv_t + \beta\mathcal{E}\, V(x_{t+1})|x_t\}$$

subject to

$$x_{t+1} = Ax_t + Bv_t + \varepsilon_{t+1}$$

with solution,

$$v_t = -Fx_t ,$$

where

$$F = \beta(Q + \beta B'PB)^{-1}B'PA ,$$

and where the P matrix solves the Ricatti equations

$$P = R + \beta A'PA - \beta^2 A'PB(Q + \beta B'PB)^{-1}B'PA .$$

Using the linear constraint, the reduced-form solution is

$$x_{t+1} = (A - BF)\,x_t + \varepsilon_{t+1} .$$

The above discussion indicates that convergence of the Ricatti equations induces an important function. This function maps the parameters of the stochastic regulator (i.e., r_{11}, r_{21}, w, q, a_{12}, b_1, and a_{22}) to the reduced-form coefficients, $A - BF$. The above discussion also reveals that iterations on the Ricatti equations amount to solving the dynamic problem 'backwards'. In the two-period reformulation of the problem, period t's value function is defined as the maximum of the current period return and the next period's expected value function. Period $t-1$'s value function is defined as the maximum of period $t-1$'s return function and period t's expected value function. Back substituting next period's value function into the current period's condition yields a sequence of optimal controls. In short, the solution procedure proceeds forward by computing past values of the optimal control.

By definition, finite time horizon problems are bounded, and their solution requires beginning in the terminal period and ending in the starting period. However, infinite time horizon problems require bounded value functions, which in turn require that distant period return functions and their control must approach zero. Notice that if $P_\circ = 0$, $F_\circ = 0$, $A - BF_\circ = 0$, and $B \neq 0$, the control in period T (i.e., v_T) is 0 as $T \to \infty$. Hence, setting $P_\circ = 0$ and iterating on

$$P_{j+1} = R + \beta A' P_j A - \beta^2 A' P_j B (Q + \beta B' P_j B)^{-1} B' P_j A$$

until the matrix P converges to a fixed point is equivalent to solving the infinite time horizon problem backwards.

The reduced-form solution of the stochastic regulator describes the movement of economic data in four different, but interrelated, dimensions. First, the reduced-form is not invariant to systematic changes in policy. A systematic change in a policy variable within the x_2 vector is represented by a change in the a_{22} coefficient. The solution procedure indicates a change in policy will not only alter the A matrix, but also will alter F, and therefore alter decision rules of agents. Hence, the problem addresses the Lucas critique of econometric policy evaluation in which reduced forms are not invariant to changes in policy.

Second, like any regression model, the reduced form coefficients measure the response of next period's state vector to a one unit change in the current state vector. Third, the reduced-form describes the response of the economy to ε_{t+1}, the vector of exogenous shocks. Specifically, such a change cannot be predicted either by agents in the model or by the econometrician, based on the current period state variables. The above setup implies a serially correlated response of the state variables to a single, uncorrelated shock. A persistently higher path of food prices following a drought describes a serially correlated response to a single, uncorrelated surprise. A bounded regulator problem implies a stable $A - BF$ matrix (one with eigenvalues less than unity in modulus). A stable $A - BF$ matrix implies the state variables can be expressed as a function of current and past shocks. In particular, let the matrix H capture the instantaneous causality (covariance) between elements of the ε_t vector, and define e_t as the vector of uncorrelated errors (Sargent, 1978). The inverted system is

$$x_{t+1} = \sum_{i=0}^{\infty} (A - BF)^i \, H e_{t-i} \, .$$

The coefficients of this impulse response function measure the contribution of past shocks on the current state vector. Equivalently, the coefficients measure the persistent movement of the state vector following a single shock.

Fourth, it can be shown that the linear (in variables) Euler equations

$$w' x_t + q u_t + \beta b_1' \mathcal{E}[r_{11} x_{1t+1} + r_{21} x_{2t+1} | x_t] = 0$$

can be factored into symmetric 'feedback' and 'feedforward' terms, and the endogenous state vector x_{1t} can be expressed as a function of the future expected stream of the exogenous state variables $\{x_{2t}\}$ (Sargent, 1987a, Ch. 14). Since $\{x_{2t+i}\}$ is assumed known, the prediction equations describing the stochastic path of x_2 are ignored. The computation of this 'perfect foresight' solution is detailed in the Appendix for the example given in a subsequent section.

The proposed estimation procedure enables the analyst to compute and make approximately correct inferences about the above responses. Successful computation permits a rigorous interpretation of the economic time series data.

3. A BOOTSTRAP ESTIMATE

The parameters of the model described in the previous section are estimated using a bootstrapping procedure and Bayes' Theorem. The prior density is an indicator function that is diffuse when the boundary conditions hold and 0 otherwise. The Bayesian bootstrap procedure permits valid inference on all of the parameters and response coefficients.

The most convenient way to explain how the bootstrap procedure is applied here is to examine the four fundamental components of the model. These are

1. Unrestricted Reduced Form

$$\begin{bmatrix} x_{1t+1} \\ x_{2t+1} \end{bmatrix} = \begin{bmatrix} \beta_{11}' & \beta_{12}' \\ 0 & \beta_{21}' \end{bmatrix} \begin{bmatrix} x_{1t} \\ x_{2t} \end{bmatrix} + \begin{bmatrix} \varepsilon_{0t+1} \\ \varepsilon_{t+1} \end{bmatrix} .$$

2. GMM estimates of the Euler equation parameters

$$d = \text{argmin}_\theta \; S(\theta, \hat{V}) \, ,$$

where

$$S(\theta, \hat{V}) = [n m_n(\theta, x)]' \, \hat{V}^{-1} [n m_n(\theta, x)]$$

and

$$m_n(\theta, x) = n^{-1} \sum_{t=1}^{\infty} e_t(\theta) \otimes z_t \,,$$

$$wx_t + qu_t + \beta b_1'[r_{11}x_{1t+1} + r_{21}x_{2t+1}] = e(\theta)\,.$$

3. Constraints

$$P_{j+1} = R + \beta A'P_jA - \beta^2 A'P_jB(Q + \beta B'P_jB)^{-1}B'P_jA$$

4. Restricted Reduced Form

$$\begin{bmatrix} x_{1t+1} \\ x_{2t+1} \end{bmatrix} = (A - BF) \begin{bmatrix} x_{1t} \\ x_{2t} \end{bmatrix} + \begin{bmatrix} 0 \\ \varepsilon_{t+1} \end{bmatrix}$$

$$F = \beta(Q + \beta B'P_jB)^{-1}B'P_jA\,.$$

In the unrestricted reduced form, β_{12} is a 'free' parameter. In the stochastic regulator, β_{12} is a function of β_{11} and β_{21}. This function or restriction may be impossible to impose on an econometric reduced-form representation. Conceptually, however, both reduced forms satisfy a similar regression structure because both residuals satisfy the condition $\mathcal{E}(\varepsilon_{t+1}x_t) = 0$. Conceptually, either regression structure could be estimated using a Seemingly Unrelated Regressions (SUR) estimator. The essence of the proposed procedure is to generate bootstrap samples using the unrestricted reduced form, restrict the bootstrap estimates to satisfy boundary conditions, and compute the restricted reduced form.

GMM estimates of θ are computed from the sample Euler equations using both the original data and the bootstrap samples. Bootstrap 'T' statistics are used to make draws on the parameters β_{21} and θ from the approximate likelihood function.

The Ricatti equations are evaluated at the parameter values of the problem. Convergence within J iterations implies the boundary conditions hold, and the prior is given a value of one. $A-BF$ is computed for draws that converge. Nonconvergence implies the boundary conditions do not hold in J iterations. In this case, the prior density is assigned a value of zero.

The key to implementing the above procedure lies in drawing the parameters from the bootstrap T statistic. The problem is similar to that of Geweke (1986) who had the convenience of exact inference in a linear regression model with normally distributed error terms. There, the pivotal element is distributed as a multivariate Student-t and can be drawn from a random number generator and added to the OLS estimate to obtain parameter draws from the likelihood. Here, the bootstrap 'T' statistic may not be pivotal, but we assume it is nearly so, so that the likelihood can conveniently be factored.

Bickel and Freedman (1981) and Freedman (1981) provide the conditions under which the distribution of a bootstrap estimate approximates the distribution of the statistic – roughly, the conditional distribution of the bootstrap sample must eventually approach the distribution of the sample. When this condition holds, the conditional distribution of the bootstrap pivot approaches the distribution of the theoretical pivot.

This result is important for both frequentist and Bayesian inference. It enables frequentists to construct accurate confidence intervals when the distribution of the sample is unknown. For a Bayesian analysis, the moments of the posterior density of the parameters must be computed. The posterior density is proportional to the product of a prior density and the likelihood function. Boos and Monahan (1986) factor the likelihood function into a function of the data and a function of the theoretical pivot. This factorization is performed under the assumption that the statistic is sufficient. Bickel and Freedman's (1981) result permits Boos and Monahan (1986) to replace the unobserved pivot with the bootstrap pivot in order to approximate the posterior density.

This result is central to our method. It permits us to make draws from the support of the approximate likelihood function using bootstrap pivots. SUR estimates of the unrestricted parameter vector $\beta = [\beta'_{11}, \beta'_{12}, \beta'_{21}]'$ and GMM estimates of the parameters θ deliver the point estimates $b = [b'_{11}, b'_{12}, b'_{21}]'$ and the point estimate d. In addition each estimator provides the covariance matrices C_b and C_d. The theoretical pivot for parameter β is $T_1 = C_b^{-1/2}(b - \beta)$, and the bootstrap pivot is $T_1^* = C_b^{*-1/2}(b^* - b)$. Since the distribution of T_1 is near T_1^*, set T_1 equal to T_1^* and

$$\beta = b - C_b^{1/2} C_b^{-1/2^*} (b^* - b) \, .$$

Repeating the same procedure for θ gives

$$\theta = d - C_d^{1/2} C_d^{-1/2^*} (d^* - d) \, .$$

The subvector $(\beta'_{21} : \theta')'$ is used to construct the matrices of the stochastic regulator problem and the Ricatti equations. For a draw in which the Ricatti equations converge, the restricted response coefficient $A - BF$ is computed. Means and standard deviations are then computed for these 'successful' draws. We illustrate this method in the next section.

4. AN EXAMPLE

One statistic of interest to agricultural economists is the food price margin. The food price margin is the difference between the value of a particular food item and the price paid to farmers for the farm component of the good. Hence, the food price margin defines the value added to the item by the processing sector. Empirical research in this area attempts to predict how food price margins change in response to a variety of exogenous shifters. Wohlgenant (1989) recognizes that nonfarm and farm factors of production are substitutes in the manufacture of food and explores the implications of input substitution for the movement of food price margins. Estimates of the parameters are obtained from functions derived from static duality theory. An earlier study, Wohlgenant (1985) explores the movement of food price margins over time. Estimates of the parameters of a univariate dynamic and stochastic optimization problem are computed. The problem illustrated in this section shows that multivariate relationships among factors of production need not be sacrificed to obtain parameter estimates of a dynamic and stochastic economic model.

Our example has the following specification:

The representative food processor's objective function

$$V^{(1)} = \max_{\{lab_t, far_t, ene_t\}} \mathcal{E}_0 \sum_{t=0}^{\infty} \beta^t \pi_t^{(1)} ,$$

where

$$\pi_t^{(1)} = p_t(\alpha_1, \alpha_2, \alpha_3) \begin{bmatrix} lab_t \\ far_t \\ ene_t \end{bmatrix} - (wag_t, r_t, enpr_t) \begin{bmatrix} lab_t \\ far_t \\ ene_t \end{bmatrix}$$

$$-(1/2)\,(lab_t, far_t, ene_t) \begin{bmatrix} h_{11} & 0 & 0 \\ 0 & 0 & 0 \\ 0 & 0 & h_{33} \end{bmatrix} \begin{pmatrix} lab_t \\ far_t \\ ene_t \end{pmatrix}$$

$$-(1/2)\,(\Delta\, lab_t, \Delta\, far_t, \Delta ene_t) \begin{bmatrix} d_{11} & d_{12} & 0 \\ d_{12} & 1 & 0 \\ 0 & 0 & d_{33} \end{bmatrix} \begin{pmatrix} \Delta\, lab_t \\ \Delta\, far_t \\ \Delta\, ene_t \end{pmatrix} ,$$

The representative farm firm's objective function

$$V^{(2)} = \max_{\{far_t\}} \mathcal{E}_0 \sum_{t=0}^{\infty} \beta^t \pi_t^{(2)} ,$$

where

$$\pi_t^{(2)} = r_t\, far_t - (c/2)\,(\Delta\, far_t)^2 .$$

The demand function for food

$$p_t = -A_1(\alpha_1, \alpha_2, \alpha_3) \begin{pmatrix} lab_t \\ far_t \\ ene_t \end{pmatrix} + dem_t .$$

The stochastic equations of motion

$$\begin{pmatrix} wag_{t+1} \\ enpr_{t+1} \\ dem_{t+1} \end{pmatrix} = \begin{bmatrix} a_{11} & 0 & 0 \\ 0 & a_{21} & 0 \\ 0 & 0 & a_{31} \end{bmatrix} \begin{pmatrix} wag_t \\ enpr_t \\ dem_t \end{pmatrix} + \begin{bmatrix} a_{12} & 0 & 0 \\ 0 & a_{22} & 0 \\ 0 & 0 & a_{32} \end{bmatrix} + \begin{pmatrix} \varepsilon_{1t+1} \\ \varepsilon_{2t+1} \\ \varepsilon_{3t+1} \end{pmatrix} .$$

The decision rule

$$\begin{pmatrix} lab_t \\ far_t \\ ene_t \end{pmatrix} = \begin{bmatrix} \rho_{11} & \rho_{12} & \rho_{13} \\ \rho_{21} & \rho_{22} & \rho_{23} \\ \rho_{31} & \rho_{32} & \rho_{33} \end{bmatrix} \begin{pmatrix} lab_{t-1} \\ far_{t-1} \\ ene_{t-1} \end{pmatrix}$$

$$+\begin{bmatrix} \rho_{14} & \rho_{15} & \rho_{16} & \rho_{17} & \rho_{18} & \rho_{19} \\ \rho_{24} & \rho_{25} & \rho_{26} & \rho_{27} & \rho_{28} & \rho_{29} \\ \rho_{34} & \rho_{35} & \rho_{36} & \rho_{37} & \rho_{38} & \rho_{39} \end{bmatrix} \begin{pmatrix} wag_t \\ wag_{t-1} \\ enpr_t \\ enpr_{t-1} \\ dem_t \\ dem_{t-1} \end{pmatrix}.$$

The price margin

$$\begin{pmatrix} p_t \\ r_t \end{pmatrix} = \begin{bmatrix} \omega_{11} & \omega_{12} & \omega_{13} \\ \omega_{21} & \omega_{22} & \omega_{23} \end{bmatrix} \begin{pmatrix} lab_{t-1} \\ far_{t-1} \\ ene_{t-1} \end{pmatrix}$$

$$+\begin{bmatrix} \omega_{14} & \omega_{15} & \omega_{16} & \omega_{17} & \omega_{18} & \omega_{19} \\ \omega_{24} & \omega_{25} & \omega_{26} & \omega_{27} & \omega_{28} & \omega_{29} \end{bmatrix} \begin{pmatrix} wag_t \\ wag_{t-1} \\ enpr_t \\ enpr_{t-1} \\ dem_t \\ dem_{t-1} \end{pmatrix}.$$

The model describes a typical food processing firm. This firm employs labor (lab), farm (far), and energy (ene) in the production of food. The firm's production process is described by a linear production function. Each period the processing firm receives the price of food (p), and pays wages (wag), farm price (r), and energy price ($enpr$). The model also describes a typical farm supplier. This supplier receives the price r for the farm inputs sold to the processor.

The processing firm incurs two types of internal capital costs associated with utilizing the three factors. First, it incurs a long-run-returns-to-scale cost associated with combining capital and the three factors. Returns-to-scale-cost parameters are embedded in the H matrix (with elements h_{ij}). Wohlgenant (1989) could not reject constant returns to scale for most of the food processing industries. We impose this restriction with $h_{22} = 0$. Second, the processing firm incurs short-run capital costs of adjustment, whose parameters are embedded in the D matrix (having elements d_{ij}). While the farm firm experiences long-run constant returns to scale, capital costs associated with output adjustments are captured in the parameter c.

The processing industry aggregate faces a consumer demand function for food output as well as the cost function of the farm sector. The variable dem represents the stochastic shifts in consumer demand, and A_1 represents the slope of the inverted demand function.[1] At the beginning of each period, a shock occurs to wages, energy prices, and demand shifts. These shocks define a set of Markov processes described by three linear difference equations. A change in the parameters of these difference equations represents a change in economic policy. The problem is to find the sequences of labor, farm, and energy that maximizes the expected social welfare

[1] The A_1 and α_i parameters are obtained or derived from previous empirical studies [Huang (1988), Putnam (1989)]. Using the sample means of the data, the demand shifter, dem, is evaluated as the residual of the consumer demand function.

function. In turn, this solution implies a sequence of equilibrium food and farm price sequences.

We used quarterly, U.S. beef industry data from 1965.1 to 1988.4 to construct the variable sequences of the model. Data sources and a description of the variable construction are available from the authors upon request. Two observations are lost to lags in the model. Four observations are lost to fourth differences. Hence, 90 observations are used in the estimation. Bootstrap samples of size 90 are drawn.

Aggregating the representative processor and the representative farm supplier's objective function gives the following dynamic programming problem

$$V^{(3)} = \max_{\{lab_t, far_t, ene_t\}} \mathcal{E}_0 \sum_{t=0}^{\infty} \beta^t \pi_t^{(3)} ,$$

where

$$\pi_t^{(3)} =$$

$$p_t(\alpha_1, \alpha_2, \alpha_3) \begin{bmatrix} lab_t \\ far_t \\ ene_t \end{bmatrix} - (wag_t, 0, enpr_t) \begin{bmatrix} lab_t \\ far_t \\ ene_t \end{bmatrix}$$

$$-(1/2)\,(lab_t, far_t, ene_t) \begin{bmatrix} h_{11} & 0 & 0 \\ 0 & 0 & 0 \\ 0 & 0 & h_{33} \end{bmatrix} \begin{pmatrix} lab_t \\ far_t \\ ene_t \end{pmatrix}$$

$$-(1/2)\,(\Delta\, lab_t, \Delta\, far_t, \Delta\, ene_t) \begin{bmatrix} d_{11} & d_{12} & 0 \\ d_{12} & 1+c & 0 \\ 0 & 0 & d_{33} \end{bmatrix} \begin{pmatrix} \Delta\, lab_t \\ \Delta\, far_t \\ \Delta\, ene_t \end{pmatrix} ,$$

subject to the stochastic equations of motion described above. The parameters of the Euler equations for this problem are estimated using GMM. Specifically, the instrumental-variable vector used to obtain the GMM estimates is $z_t = [lab_{t-1}, far_{t-1}, ene_{t-1}, wag_{t-1}, enpr_{t-1}, p_{t-1}, dem_{t-1}]'$. Cholesky decompositions of the SUR and GMM estimates of the covariance matrices are computed to form the 'T' statistics.

The equilibrium of the model is found by solving the following dynamic programming problem:

$$V^{(4)} = \max_{\{lab_t, far_t, ene_t\}} \mathcal{E}_0 \sum_{t=0}^{\infty} \beta^t \pi_t^{(4)} ,$$

where

$$\pi_t^{(4)} =$$

$$-(wag_t, enpr_t, dem_t) \begin{bmatrix} 1 & 0 & 0 \\ 0 & 1 & 0 \\ -\alpha_1 & -\alpha_2 & -\alpha_3 \end{bmatrix} \begin{bmatrix} lab_t \\ far_t \\ ene_t \end{bmatrix}$$

$$-(lab_t, far_t, ene_t)\,[H + A_1\alpha\alpha']\begin{pmatrix} lab_t \\ far_t \\ ene_t \end{pmatrix}$$

$$-(1/2)\,(\Delta\, lab_t, \Delta\, far_t, \Delta\, ene_t)\begin{bmatrix} d_{11} & d_{12} & 0 \\ d_{12} & 1+c & 0 \\ 0 & 0 & d_{33} \end{bmatrix}\begin{pmatrix} \Delta\, lab_t \\ \Delta\, far_t \\ \Delta\, ene_t \end{pmatrix},$$

subject to the equations of motion given above. We compute the posterior distribution of the objective function parameters and the linear stochastic difference equations. The prior is assigned a value 1 if the Ricatti equations associated with $V^{(4)}$ converge within 150 iterations. Otherwise, the prior is assigned a value zero. 664 of 1000 draws from the bootstrap likelihood resulted in convergent Ricatti equations.

We also compute the posterior for $A - BF$. $A - BF$ represent the response coefficients of the reduced-form input demand functions. Combining $A - BF$ with the consumer demand function gives the parameters of the food price equation. Combining $A - BF$ with the farm supplier's Euler equations gives the parameters of the farm price equation. The food price and the farm price functions constitute the food price margin function.

In Tables 1 to 3, we report the means and standard deviations (in parentheses) of the posterior distribution. We assume a quadratic loss function. Therefore, the mean represents our parameter estimate because it minimizes the loss function (Zellner, 1987). The standard error serves as the measure of dispersion of the posterior.

Table 1 reports the estimates of the parameters of the stochastic regulator, its reduced-form solution, and the price margin functions. The negative estimate of h_{11} suggests that labor is a capital saving input in the long-run in the beef industry. The results also suggest firms consume capital when they adjust labor. ($d_{11} > 0$) However, they can offset capital adjustment costs by substituting farm inputs for labor ($d_{12} < 0$). Our estimate of the parameter c (62.6) indicates the short-run supply of farm inputs facing the processing industry is upward sloping.

Table 1 also reports the parameter estimates of the equation of motion. The results indicate the demand shifter displays oscillating (complex roots) patterns. The average period from peak-to-peak is approximately one month (about $1/3$ of a quarter). Also, the results indicate that changes in energy prices have been more permanent than have changes in wages.

Estimates of the coefficients of the equilibrium input demand functions are reported in Table 1. Coefficient estimates of the reduced form are composite functions of all or many of the parameters of the problem. Hence, the standard deviations associated with the composite coefficients embody the standard deviations of many parameters. The composite coefficients sometimes capture opposite effects. We estimate a negative steady-state cost of capital associated with labor. We also estimate a positive dynamic cost of capital associated with labor ($h_{11} < 0$, $d_{11} > 0$). The result of these offsetting effects is a positive response of labor to current period wages (0.173). Apparently, it is the negative steady-state costs that induce firms to hire less labor when consumer demand increases (-0.111). Our results are consistent with

TABLE 1

Parameter estimates, beef model.*

The representative food processor's objective function

$$V^{(1)} = \max_{\{lab_t, far_t, ene_t\}} \mathcal{E}_0 \sum_{t=0}^{\infty} (.95)^t \pi_t^{(1)},$$

where

$$\pi_t^{(1)} = p_t(.642,\ .730,\ .508) \begin{bmatrix} lab_t \\ far_t \\ ene_t \end{bmatrix}$$

$$-(wag_t, r_t, enpr_t) \begin{bmatrix} lab_t \\ far_t \\ ene_t \end{bmatrix} - (1/2)\,(lab_t, far_t, ene_t) \begin{bmatrix} -10.2 & .000 & .000 \\ (8.1) & & \\ .000 & .000 & .000 \\ .000 & .000 & 6.34 \\ & & (5.3) \end{bmatrix} \begin{pmatrix} lab_t \\ far_t \\ ene_t \end{pmatrix}$$

$$-(1/2)\,(\Delta\, lab_t, \Delta\, far_t, \Delta\, ene_t) \begin{bmatrix} 32.1 & -48.0 & .000 \\ (29.2) & (43.3) & \\ -48.0 & 1.00 & .000 \\ (43.3) & & \\ .000 & .000 & 4.72 \\ & & (4.5) \end{bmatrix} \begin{pmatrix} \Delta\, lab_t \\ \Delta\, far_t \\ \Delta\, ene_t \end{pmatrix}.$$

The representative farm firm's objective function

$$V^{(2)} = \max_{\{far_t\}} \mathcal{E}_0 \sum_{t=0}^{\infty} (.95)^t \pi_t^{(2)}$$

where,

$$\pi_t^{(2)} = r_t\, far_t - (1/2)\, \underset{(43.3)}{62.6} (far_t - far_{t-1})^2.$$

The demand function for food

$$p_t = -.563(.642,\ .730,\ .508) \begin{pmatrix} lab_t \\ far_t \\ ene_t \end{pmatrix} + dem_t.$$

The stochastic equations of motion

$$\begin{pmatrix} wag_{t+1} \\ enpr_{t+1} \\ dem_{t+1} \end{pmatrix} = \begin{bmatrix} 1.09 & .000 & .000 \\ (.10) & & \\ .000 & .953 & .000 \\ & (.12) & \\ .000 & .000 & .995 \\ & & (.12) \end{bmatrix} \begin{pmatrix} wag_t \\ enpr_t \\ dem_t \end{pmatrix}$$

$$+ \begin{bmatrix} -.21 & .000 & .000 \\ (.09) & & \\ .000 & -.037 & .000 \\ & (.10) & \\ .000 & .000 & -.286 \\ & & (.10) \end{bmatrix} + \begin{pmatrix} \varepsilon_{1t+1} \\ \varepsilon_{2t+1} \\ \varepsilon_{3t+1} \end{pmatrix}.$$

Table 1 (continued)

The decision rule

$$\begin{pmatrix} lab_t \\ far_t \\ ene_t \end{pmatrix} = \begin{bmatrix} .268 & .042 & .007 \\ (.30) & (.09) & (.02) \\ & & \\ -.39 & .915 & .003 \\ (.28) & (.21) & (.02) \\ & & \\ .002 & -.038 & .292 \\ (.07 & (.12) & (.19) \end{bmatrix} \begin{pmatrix} lab_{t-1} \\ far_{t-1} \\ ene_{t-1} \end{pmatrix}$$

$$+ \begin{bmatrix} .173 & -.011 & -.013 & .000 & -.111 & .008 \\ (.49) & (.05) & (.05) & (.00) & (.18) & (.03) \\ & & & & & \\ .014 & .001 & .012 & -.000 & .007 & -.009 \\ (.22) & (.02) & (.11) & (.01) & (.12) & (.02) \\ & & & & & \\ -.014 & .001 & -.197 & -.002 & .104 & -.007 \\ (.06) & (.01) & (.55) & (.05) & (.22) & (.05) \end{bmatrix} \begin{pmatrix} wag_t \\ wag_{t-1} \\ enpr_t \\ enpr_{t-1} \\ dem_t \\ dem_{t-1} \end{pmatrix}.$$

The price margin

$$\begin{pmatrix} p_t \\ r_t \end{pmatrix} = \begin{bmatrix} -.063 & .380 & .087 \\ (.18) & (.09) & (.06) \\ & & \\ -3.86 & -.053 & -.00 \\ (12.5) & (.51) & (2.0) \end{bmatrix} \begin{pmatrix} lab_{t-1} \\ far_{t-1} \\ ene_{t-1} \end{pmatrix}$$

$$+ \begin{bmatrix} .064 & -.003 & -.056 & -.001 & .992 & -.003 \\ (.11) & (.01) & (.14) & (.01) & (.10) & (.02) \\ & & & & & \\ .587 & -.365 & .009 & .006 & -.364 & .509 \\ (1.0) & (.67) & (.16) & (.12) & (1.2) & (.88) \end{bmatrix} \begin{pmatrix} wag_t \\ wag_{t-1} \\ enpr_t \\ enpr_{t-1} \\ dem_t \\ dem_{t-1} \end{pmatrix}.$$

* Reported values are means of the posterior, and the numbers in parentheses are standard errors of the posterior.

the notion that consumers have shifted toward products containing more nonfarm inputs.

These results are also used to trace the impacts of exogenous changes on the food price margin. Our results indicate that a wage increase induces an increase in labor demand. However, positive adjustment costs associated with labor dampen this increase. In response to a wage increase, firms substitute farm inputs for labor. This raises the demand for farm inputs and increases farm prices. Our results suggest the increase in wages raises the marginal costs of processing. However, the larger increase in farm prices narrows the food price margin. The results also suggest a weak relationship between the demand for farm inputs and a (positive) shift in consumer demand. Our point estimate is slightly negative (-0.007). Hence, we estimate that the price margin widens when consumers increase their demand for beef.

TABLE 2
Estimates of the perfect foresight solution, beef prices.*

	Food Price$_t$, p_t			Farm Price$_t$, r_t		
j	$\mathcal{E}_t\, wag_{t+j}$	$\mathcal{E}_t\, enpr_{t+j}$	$\mathcal{E}_t\, dem_{t+j}$	$\mathcal{E}_t\, wag_{t+j}$	$\mathcal{E}_t\, enpr_{t+j}$	$\mathcal{E}_t\, dem_{t+j}$
1	–0.0489	–0.0084	1.0136	3.5922	1.9117	–3.6690
	(.109)	(.056)	(.082)	(5.38)	(3.18)	(5.42)
2	–0.0030	0.0071	–0.0073	–1.6438	–1.1511	1.8734
	(.040)	(.019)	(.033)	(4.13)	(2.49)	(4.23)
3	–0.0049	0.0063	–0.0028	0.1807	0.0703	–0.1549
	(.018)	(.011)	(.017)	(2.13)	(1.23)	(2.16)
4	0.0001	0.0069	–0.0054	–0.2107	–0.1368	0.2189
	(.011)	(.007)	(.012)	(1.36)	(.746)	(1.39)
5	–0.008	0.0059	–0.0038	0.0691	0.0339	–0.0575
	(.007)	(.006)	(.008)	(.987)	(.535)	(1.06)
6	0.0004	0.0054	–0.0042	–0.0590	–0.0333	0.0512
	(.005)	(.004)	(.007)	(.755)	(.409)	(.860)
7	0.0000	0.0047	–0.0033	0.0276	0.0153	–0.0206
	(.003)	(.004)	(.006)	(.602)	(.329)	(.714)
8	0.0004	0.0041	–0.0032	–0.0210	–0.0095	0.0136
	(.003)	(.003)	(.005)	(.487)	(.270)	(.592)

* Reported values are means of the posterior. The values in parentheses are standard errors of the posterior.

Eckstein (1985) demonstrates that a univariate stochastic regulator problem can be used to compute a useful variety of response elasticities. The results presented in Tables 2 and 3 illustrate that the above procedure is well-suited to the statistical estimation of such responses implied by a more general problem. Table 2 reports the estimates of the perfect foresight solution. This solution gives the current period response to a known or expected change that occurs j periods into the future. The idea is that firms adjust production in the current period to reduce adjustment costs later.

The responses reported in Table 2 display the small effect that future wage increases exert on current period food prices. This small response is partly due to the offsetting static and dynamic costs of adjusting labor. Table 2 also reports that future wage changes exert a positive effect on farm price. Evidently, firms substitute

TABLE 3
Impulse response estimates, beef model.*

	Food Price$_t$, p_t			Farm Price$_t$, r_t		
j	wag_{t-j}	$enpr_{t-j}$	dem_{t-j}	wag_{t-j}	$enpr_{t-j}$	dem_{t-j}
0	0.0000	0.0000	0.0000	2.1067	–0.0265	-2.2515
	0.0000	0.0000	0.0000	(2.41)	(.736)	(2.60)
1	0.0643	–0.0559	0.9935	0.5871	0.0082	–0.3858
	(.106)	(.136)	(.103)	(1.05)	(.178)	(1.20)
2	0.0750	–0.0635	1.0034	0.2701	0.0121	0.1508
	(.118)	(.150)	(.148)	(.568)	(.147)	(.573)
3	0.0722	–0.0609	0.7337	–0.0097	0.0269	0.4512
	(.118)	(.151)	(.168)	(.386)	(.132)	(.444)
4	0.0623	–0.0549	0.4591	–0.0452	0.0205	0.3978
	(.109)	(.143)	(.183)	(.266)	(.120)	(.389)
5	0.0520	–0.0485	0.2620	–0.0962	0.0273	0.3114
	(.095)	(.130)	(.191)	(.216)	(.105)	(.339)
6	0.0419	–0.0425	0.1471	–0.0778	0.0189	0.1830
	(.081)	(.114)	(.179)	(.181)	(.093)	(.281)
7	0.0334	–0.0372	0.0906	–0.0810	0.0220	0.1010
	(.068)	(.098)	(.152)	(.156)	(.079)	(.236)

* Reported values are means of the posterior. The values in parentheses are standard errors of the posterior.

farm inputs for labor before an expected wage increase. Our results also indicate the price and quantity demanded of farm commodities change before a known increase in energy price.

The impulse response coefficients are reported in Table 3. Our results measure the change in food and farm prices following a shock to an exogenous variable. These coefficients account for the contemporaneous relationships among the various shocks. Hence, it is difficult to provide an intuitive explanation of the results presented in Table 3.

5. CONCLUSIONS

This study uses the Bayesian bootstrap to compute econometric estimates of stochastic, dynamic programming problems. Typically, statistical inferences on the reduced-form coefficients of such a problem are difficult because of the complex cross-equation restrictions that characterize such solutions. Likewise, direct estimation of the problem's parameters requires the estimates to adhere to boundary conditions, which when imposed, require classical techniques to be significantly modified or discarded (since the boundary condition cannot be checked by evaluating, for example, the eigenvalues of a matrix). By contrast, our procedure combines textbook algorithms with the Bayesian bootstrap to form an estimator that is well suited to impose such a restriction.

The estimator holds value for analysts facing difficulties imposing restrictions on any econometric model. It also should be useful for analysts pursing a Bayesian analysis, but who are uncomfortable with the usual assumption of normally distributed error terms. All that is required is a statistical representation from which the analyst can draw bootstrap samples of the variables of the model. The estimator could be used, for example, to estimate static duality models when one is concerned with imposing the required curvature restrictions.

REFERENCES

Baxter, M., M.J. Cricini, and K.G. Rouwenhorst: 1990, 'Solving the stochastic growth model by a discrete state-space, euler-equation approach', *Journal of Business and Economic Statistics* 8, 19–21.

Bickel, P.J., and D.A. Freedman: 1981, 'Some asymptotic theory for the bootstrap', *The Annals of Statistics* 9, 1196–1217.

Boos D.D., and J.F. Monahan: 1986, 'Bootstrap methods using prior information', *Biometrika* 73, 77–83.

Christiano, L.J.: 1990, 'Solving the stochastic growth model by linear-quadratic approximation and by value-function iteration', *Journal of Business and Economic Statistics* 8, 23–26.

Coleman, W.J.: 1990, 'Solving the stochastic growth model by policy-function iteration', *Journal of Business and Economic Statistics* 8, 27–29.

den Haan, W.J., and A. Marcet: 1990, 'Solving the stochastic growth model by parameterizing expectations', *Journal of Business and Economic Statistics* 8, 31–34.

Eckstein, Z.: 1985, 'The dynamics of agriculture supply: a reconsideration', *American Journal of Agricultural Economics* 67, 204–214.

Freedman, D.A.: 1981, 'Bootstrapping regression models', *The Annals of Statistics* 9, 1218–1228.

Gagnon, J.E.: 1990, 'Solving the stochastic growth model by deterministic extended path', *Journal of Business and Economic Statistics* 8, 35–38.

Gallant, A.R.: 1987, *Nonlinear Statistical Models*, New York: John Wiley and Sons.

Gallant, A.R., and G.H. Golub: 1984, 'Imposing curvature restrictions on flexible functional forms', *Journal of Econometrics* 26, 295–321.

Geweke, J.: 1986, 'Exact inference in the inequality constrained normal linear regression model', *Journal of Applied Econometrics* 1, 127–141.

Huang, K.: 1988, 'An inverse demand system for U.S. composite goods', *American Journal of Agricultural Economics* 70, 902–909.

Labadie, P.: 1990, 'Solving the stochastic growth model by using a recursive mapping based on least squares projection', *Journal of Business and Economic Statistics* 8, 39–40.

Lucas, R.E.: 1976, 'Econometric policy evaluation: a critique', *The Phillips Curve and the Labor Market* (K. Brunner and A. Meltzer eds) Volume 1 of Carnegie-Rochester Conferences in Public Policy, a supplementary series to the *Journal of Monetary Economics,* Amsterdam: North Holland.

McGratten, E.R.: 1990, 'Solving the stochastic growth model by linear-quadratic approximation', *Journal of Business and Economic Statistics* 8, 41–44.

Miranda, M.J., and J.W. Glauber: 1991. "Estimation of dynamic nonlinear rational expectations models of commodity markets with private and government stockholding." Paper presented at the annual meetings of the American Agricultural Economics Association. Manhattan, Kansas. August 4–7, 1991.

Putnam, J.J.: 1989, *Food Consumption, Price, and Expenditures* USDA/ERS Satistical Bulletin No. 773.

Sargent, T.J.: 1987a, *Macroeconomic Theory,* Boston: Academic Press.

Sargent, T.J.: 1987b, *Dynamic Macroeconomic Theory,* Cambridge: Harvard University Press.

Sargent, T.J.: 1978: 'Estimation of dynamic demand schedules under rational expectations', *Journal of Political Economy,* 86, 1009–1044.

Sims, C.: 1990, 'Solving the stochastic growth model by backsolving with a particular nonlinear form for the decision rule', *Journal of Business and Economic Statistics* 8, 45–48.

Tauchen, G.: 1990, 'Solving the stochastic growth model by using quadrature methods and value-function iterations', *Journal of Business and Economic Statistics* 8, 49–51.

Taylor J.B. and H. Uhlig: 1990, 'Solving nonlinear stochastic growth models: a comparison of alternative solution methods', *Journal of Business and Economic Statistics* 8, 1–17.

Wohlgenant, M.K.: 1989, 'Demand for farm output in a complete system of demand functions', *American Journal of Agricultural Economics,* 71, 241–252.

Wohlgenant, M.K.: 1985, 'Competitive storage, rational expectations, and short-run food price determination', *American Journal of Agricultural Economics,* 67, 739–748.

Zellner, A.: 1987b, *An Introduction to Bayesian Inference in Econometrics,* Malibar: Robert E. Krieger Publishing Company.

GREGORY C. CHOW

Computation of Optimum Control Functions by Lagrange Multipliers

ABSTRACT. An algorithm is proposed to compute the optimal control function without solving for the value function in the Bellman equation of dynamic programming. The method is to solve a pair of vector equations for the control variables and the Lagrange multipliers associated with a set of first-order conditions for an optimal stochastic control problem. It approximates the vector control function and the vector Lagrangean function locally for each value of the state variables by linear functions. An example illustrates that such a local approximation is better than global approximations of the value function.

Previously (Chow 1992a, 1993) I have shown that the optimum control function of a standard optimum control problem can be derived more conveniently by using Lagrange multipliers than solving the Bellman partial differential equation for the value function. This derivation also provides numerical methods for computing the value of the optimum control corresponding to a given value of the state variable that are more accurate than those based on solving the Bellman equation. This paper explains the gain in numerical accuracy and illustrates it by example.

1. DERIVATION OF THE OPTIMAL CONTROL FUNCTION

Consider the following standard optimum control problem in discrete time (an analogous problem in continuous time is considered in Chow (1993), and the results of this paper apply equally well to that problem). Let x_t be a column vector of p state variables and u_t be a vector of q control variables. Let r be a concave and twice differentiable function and β be a discount factor. E_t denotes conditional expectation given information at time t, which includes x_t. The problem is

$$\max_{\{u_t\}_{t=0}^{\infty}} \sum_{t=0}^{\infty} E_t\, \beta^t\, r(x_t, u_t) \tag{1}$$

subject to

$$x_{t+1} = f(x_t, u_t) + \varepsilon_{t+1}\,, \tag{2}$$

where ε_{t+1} is an i.i.d. random vector with mean zero and covariance matrix Σ. Chow (1992a) solves this problem by introducing the $p \times 1$ vector λ_t of Lagrange multipliers and setting to zero the derivatives of the Lagrangean expression

D. A. Belsley (ed.), Computational Techniques for Econometrics and Economic Analysis, 65–72.

$$\mathcal{L} = \sum_{t=0}^{\infty} E_t\{\beta^t r(x_t, u_t) + \beta^{t+1}\lambda'_{t+1}[x_{t+1} - f(x_t, u_t) - \varepsilon_{t+1}]\} \tag{3}$$

with respect to u_t and x_t $(t = 0, 1, 2, \ldots)$. The first-order conditions are

$$\frac{\partial}{\partial u_t} r(x_t, u_t) + \beta \frac{\partial}{\partial u_t} f'(x_t, u_t) E_t\lambda_{t+1} = 0\,, \tag{4}$$

$$\lambda_t = \frac{\partial}{\partial x_t} r(s_t, u_t) + \beta \frac{\partial}{\partial x_t} f'(x_t, u_t) E_t\lambda_{t+1}\,. \tag{5}$$

The optimum control at time t is obtained by solving equations (4) and (5) for u_t and λ_t.

The difficult part in solving these equations is the evaluation of the conditional expectation $E_t\lambda_{t+1}$, a problem to be treated shortly. We first point out the main differences between this approach and that of solving the Bellman partial differential equation for the value function $V(x)$. First, it is not necessary to know the value function to derive the optimum control function since the latter is a functional, not of V, but of the vector λ of derivatives of V with respect to the state variables. Thus, obtaining the value function V requires more than is needed to obtain the optimum control function and hence solves a more difficult problem than necessary. For example, in the problem of static demand theory derived from maximizing consumer utility subject to a budget constraint, Bellman's method amounts to finding the indirect utility function by solving a partial differential equation, whereas we would apply the method of Lagrange multipliers in obtaining the demand function. Second, our equation (5) could be obtained by differentiating the Bellman equation with respect to the state variables. This is a very important first order condition for optimality, but it is ignored when one tries to solve the Bellman equation for the value function and thus makes the solution of the optimum control problem more difficult. Third, for most realistic applied problems an analytical solution for the value function is not available. A common practice when solving the Bellman equation is to use a global approximation to the value function when deriving the optimum control function. By contrast, in solving equations (4) and (5) for a given x_t we avoid using a global approximation to the Lagrange function in the neighborhood of x_t and use instead a linear function to approximate λ locally for each x_t. This typically yields a more accurate approximation to the Lagrange function and hence to the corresponding value function in the Bellman approach.

2. NUMERICAL SOLUTION OF THE FIRST ORDER CONDITIONS

To provide a numerical method for solving the first order conditions (4) and (5), we approximate the Lagrange function in the neighborhood of x_t by a linear function,

$$\lambda_t = H_t x_t + h_t\,, \tag{6}$$

where the t subscripts of the parameters H_t and h_t indicate that the linear function (6) applies to points not too far from x_t, in particular to x_{t+1}. Thus

$$\lambda_{t+1} = \lambda_t(x_{t+1}) = H_t x_{t+1} + h_t = H_t f(x_t, u_t) + H_t \varepsilon_{t+1} + h_t \, ,$$
$$E_t \;\; \lambda_{t+1} = H_t f(x_t, u_t) + h_t \, . \tag{7}$$

Taking x_t as given, we try to solve (4) and (5) for u_t and λ_t using (7) for $E_t\lambda_{t+1}$. Substituting (7) into (4) yields

$$\frac{\partial r}{\partial u_t} + \beta \, \frac{\partial f'}{\partial u_t} \, (H_t f + h_t) = 0 \, . \tag{8}$$

Assuming tentatively H_t and h_t to be known, we solve (8) for u_t using linear approximations of $\frac{\partial r}{\partial u_t}$ and f

$$\frac{\partial r}{\partial u_t} = K_{2t} u_t + K_{21t} x_t + k_{2t} \, ,$$
$$\frac{\partial r}{\partial x_t} = K_{1t} x_t + K_{12t} u_t + k_{1t} \, , \tag{9}$$

$$f(x_t, u_t) = A_t x_t + C_t u_t + b_t \, , \tag{10}$$

where the time subscripts for the parameters of the linear functions indicate that the functions are valid for values of x and u near x_t and the optimal u_t^*. These parameters are obtained by evaluating the partial derivatives of $\frac{\partial r}{\partial u}$ and f at x_t and some initial value for u_t^*, the latter to be revised after each iteration. Substituting (9) and (10) into (8) gives

$$K_{2t} u_t + K_{21t} x_t + k_{2t} + \beta C_t' H_t (A_t x_t + C_t u_t + b_t) + \beta C_t' h_t = 0 \, . \tag{11}$$

Equation (11) can be solved for u_t, yielding

$$u_t = G_t x_t + g_t \, , \tag{12}$$

where

$$G_t = -(K_{2t} + \beta C_t' H_t C_t)^{-1} \, (K_{21t} + \beta C_t' H_t A_t) \, , \tag{13}$$

$$g_t = -(K_{2t} + \beta C_t' H_t C_t)^{-1} \, [k_{2t} + \beta C_t' (H_t b_t + h_t)] \, . \tag{14}$$

To find the parameters H_t and h_t for λ_t, we substitute (6), (7), (9), (10) and (12) into (5) to get

$$\begin{aligned} H_t x_t + h_t = \;& K_{1t} x_t + K_{12t}(G_t x_t + g_t) + k_{1t} \\ & \beta A_t' H_t (A_t x_t + C_t G_t x_t + C_t g_t + b_t) + \beta A_t' h_t \, . \end{aligned} \tag{15}$$

Equating coefficients of (15) yields

$$H_t = K_{1t} + K_{12t} G_t + \beta A_t' H_t (A_t + C_t G_t) \, , \tag{16}$$

$$h_t = (K_{12t} + \beta A_t' H_t C_t)\, g_t + k_{1t} + \beta A_t'(H_t b_t + h_t)\,. \tag{17}$$

To solve equations (4) and (5) numerically, we assume some initial value for the optimal u_t^* and linearize $\partial r/\partial u_t$, $\partial r/\partial x_t$ and f about x_t and this value of u_t^* as in (9) and (10). We then solve the pair of equations (13) and (16) iteratively for G_t and H_t. Given G_t and H_t, the pair of equations (14) and (17) can be solved iteratively for g_t and h_t. The value of optimal control u_t^* is found by $G_t x_t + g_t$. This value will be used to relinearize $\partial r/\partial u$, $\partial r/\partial x$ and f until convergence.

The reader may recognize that the numerical method suggested in this paper amounts to solving the well known matrix Ricatti equations (13) and (16) for G_t and H_t in linear-quadratic control problems. However, there are two important differences from the standard treatment of stochastic control by dynamic programming. First, our derivation is different as it does not use the value function at all. Second, we emphasize the solution of two equations (4) and (5) for u_t and λ_t while treating x_t as given. We have avoided global approximations to the functions $u(x)$ and $\lambda(x)$ which can lead to large errors. We employ linear approximations to $u(x)$ and $\lambda(x)$ only locally about a given x_t and build up the nonlinear functions $u(x)$ and $\lambda(x)$ by these locally linear approximations for different x_t. To generalize our second point, we can choose other methods to solve equations (4) and (5) for a given x_t. We could, for example, use a quadratic approximation to $\lambda(x)$ as discussed in Chow (1992a). We leave other numerical methods for solving equations (4) and (5) for future research.

3. AN ILLUSTRATIVE EXAMPLE

To demonstrate how a nonlinear optimal control function is computed numerically by locally linear approximations, I use a baseline real business cycle model presented by King, Plosser and Rebelo (1988) and analyzed by Watson (1990). The model consists of two control variables u_{1t} and u_{2t}, representing consumption and labor input, respectively, and two state variables x_{1t} and x_{2t}, denoting, respectively, log A_t and capital stock at the beginning of period t, where A_t represents technology in the production function $q_t = x_{2t}^{1-\alpha}(A_t u_{2t})^{\alpha}$. The dynamic process (2) is

$$\begin{aligned} x_{1t} &= \gamma + x_{1,t-1} + \varepsilon_t\,, \\ x_{2t} &= (1-\delta)\, x_{2,t-1} + x_{2,t-1}^{1-\alpha}\ \exp(\alpha x_{1,t-1})\, u_{2,t-1}^{\alpha} - u_{1,t-1}\,. \end{aligned} \tag{18}$$

The first equation assumes $x_{1t} = \log A_t$ to be a random walk with a drift γ, ε_t being a random shock to technology. The second equation gives the evolution of capital stock x_{2t}, with δ denoting the rate of depreciation and investment being the difference between output q_{t-1} given by the production function and consumption $u_{1,t-1}$. The utility function r in (1) is assumed to be

$$r = \log u_{1t} + \theta\, \log(1-u_{2t})\,, \tag{19}$$

TABLE 1
Optimal control variables corresponding to selected state variables.

	u_1	u_2	x_1	x_2
1.	4.865	0.227	3.466	13.098
2.	5.028	0.219	3.518	13.423
3.	5.297	0.217	3.544	13.754
4.	5.341	0.208	3.559	14.153
5.	5.535	0.206	3.606	14.404
6.	5.656	0.202	3.663	14.672
7.	5.979	0.202	3.738	15.045
8.	6.486	0.208	3.814	15.754
9.	6.868	0.211	3.851	16.765
10.	7.232	0.212	3.851	17.783
11.	7.568	0.212	3.872	18.641
12.	7.901	0.220	3.884	19.715
13.	8.100	0.217	3.862	20.737
14.	8.681	0.230	3.891	21.636
15.	8.844	0.236	3.880	22.893
16.	8.807	0.230	3.862	24.126
17.	9.315	0.234	3.877	25.119
18.	9.983	0.245	3.895	26.441
19.	10.427	0.254	3.913	27.750

where $1 - u_{2t}$ denotes leisure.

There are five parameters in this model: α, the labor exponent in a Cobb-Douglas production function; β, the dicount factor; γ, the drift in the random walk process for x_{1t}, which is log of the Solow residual in the production function; δ, the rate of depreciation for capital stock; and θ, the weight given to leisure in the log-linear utility function of consumption u_{1t} and leisure $1 - u_{2t}$. In Chow (1992b) I have estimated these five parameters by maximum likelihood using quarterly data of the United States from 1951.1 to 1988.4, covering 38 years. The subject of statistical estimation by maximum likelihood does not concern us in this paper. Here we take the resulting set of values for the parameters as given and examine how the linear approximations to the optimal control function change with the state variables. Let $\alpha = .6368$, $\beta = .8453$, $\gamma = .00304$, $\delta = 1.77 \times 10^{-8}$, and $\theta = 3.5198$. I have computed the optimal values for the control variables corresponding to 19 sets of state variables, which are the historical values of these state variables in the first quarters of the 19 years 1951, 1953, ..., 1987. The values of the four variables are given in Table 1. The parameters G_{11}, G_{12}, G_{21}, G_{22}, g_1, g_2 of the linear function corresponding to each set of state variables are given in Table 2. Table 2 illustrates how poor a global linear approximation to the optimal control function would be, as

TABLE 2
Parameters of linear optimal control functions.

	G_{11}	G_{12}	G_{21}	G_{22}	g_1	g_2
1.	1.255	0.0546	−0.219	−0.0095	0.165	1.223
2.	1.207	0.0513	−0.213	−0.0090	0.558	1.112
3.	1.280	0.0531	−0.209	−0.0087	0.384	1.107
4.	1.169	0.0471	−0.199	−0.0080	1.059	1.055
5.	1.197	0.0474	−0.199	−0.0079	1.085	1.071
6.	1.170	0.0455	−0.198	−0.0077	1.382	1.086
7.	1.263	0.0479	−0.203	−0.0077	1.182	1.136
8.	1.470	0.0532	−0.210	−0.0076	0.517	1.195
9.	1.602	0.0545	−0.211	−0.0072	0.207	1.202
10.	1.714	0.0550	−0.206	−0.0066	−0.048	1.167
11.	1.789	0.0547	−0.202	−0.0062	−0.123	1.149
12.	2.009	0.0581	−0.207	−0.0060	−0.957	1.164
13.	1.974	0.0543	−0.196	−0.0054	−0.495	1.090
14.	2.420	0.0638	−0.205	−0.0054	−2.500	1.142
15.	2.542	0.0633	−0.204	−0.0051	−2.820	1.122
16.	2.316	0.0548	−0.193	−0.0046	−1.354	1.042
17.	2.558	0.0581	−0.192	−0.0044	−2.262	1.036
18.	3.021	0.0652	−0.195	−0.0042	−4.254	1.050
19.	3.372	0.0693	−0.198	−0.0041	−5.693	1.063

TABLE 3
Regressions of coefficients of linear control functions on state variables (t statistics in parentheses).

Dependent variables	Explanatory variables 1	x_1	x_2	R^2
G_{11}	0.391	−0.346	0.148	0.960
	(0.337)	(−0.986)	(13.16)	
G_{12}	0.044	−0.0029	0.0012	0.683
	(1.40)	(−0.304)	(3.93)	
G_{21}	−0.204	−0.0052	0.0011	0.405
	(4.21)	(−0.350)	(2.35)	
G_{22}	−0.023	0.0031	0.00027	0.975
	(−9.93)	(4.37)	(11.84)	

the parameters of the locally linear approximations change with the state variables. To describe the changes, Table 3 presents linear regressions of four parameters on the two state variables and the accompanying t statistics and R^2 for descriptive purposes only (as the regressions are not based on a stochastic model).

For this example, we set the maximum number of iterations for solving the pair of equations G_t and H_t using (13) and (16) to 25, given each value for u_t^* used in linearizing $\partial r/\partial u$, $\partial r/\partial x$ and f. For our criterion of convergence to three significant figures, the maximum number of 25 is found to be better than 50 and 20. Once the optimal linear control function is found for x_1 and x_2 as of 1951.1, the optimum u_t^*, G_t, H_t, g_t and h_t can be used as initial values to compute the optimal linear control function corresponding to x_1 and x_2 as of 1953.1, and so forth. It takes about eight hours on a 486 personal computer to maximize a likelihood function with respect to the five parameters using a simulated annealing maximization algorithm (see Goffe, Ferrier and Rogers, 1992) which evaluates the likelihood function about 14,000 times (or about two seconds per evaluation of the likelihood function). At each evaluation, one must find the linear optimal control function for the given parameters, compute the residuals of the observed values of the control variables from the computed optimal values for 152 quarters, compute the value of the likelihood function, and determine the new values of the five parameters for the next functional evaluation, which may be time consuming. Hence, merely computing the linear optimal control function for a given set of parameters in our example should take less than one second of a 486 computer using Gauss.

In this paper, I have shown how locally linear optimal control functions can be computed for a standard stochastic control problem in discrete time. The algorithm is based on solving two equations for the vectors of control variables and Lagrange multipliers, given the vector of state variables. It is easy to implement using a personal computer. It can serve as an important component of an algorithm for the statistical estimation of the parameters of a stochastic control problem in econometrics.

ACKNOWLEDGEMENTS

The author would like to thank Chunsheng Zhou for excellent programming assistance in obtaining the numerical results reported in this paper and David Belsley for helpful comments on an early draft.

REFERENCES

Chow, Gregory C., "Dynamic optimization without dynamic programming," *Economic Modelling*, 9 (1992a), 3–9.

Chow, Gregory C., "Statistical estimation and testing of a real business cycle model," Princeton University, Econometric Research Program, Research Memorandum No. 365 (1992b)

Chow, Gregory C.,"Optimal control without solving the Bellman equation," *Journal of Economic Dynamics and Control, 17* (1993).

Goffe, William L., Gary Ferrier and John Rogers, "Global optimization of statistical functions," in *Computational Economics and Econometrics, Vol. 1*, eds. Hans M. Amman, D. A. Belsley, and Louis F. Pau, Dordrecht: Kluwer, 1992.

King, Robert G., Charles I. Plosser and S. T. Rebelo, "Production, growth, and business cycles: II. New Directions," *Journal of Monetary Economics*, 21 (1988), 309–342.

Watson, Mark W., "Measures of fit for calibrated models," Northwestern University and Federal Reserve Bank of Chicago, mimeo, 1990.

PART TWO

The Computer and Economic Analysis

DAVID KENDRICK

Computational Approaches to Learning with Control Theory

ABSTRACT. Macroeconomics has just passed through a period in which it was assumed that everyone knew everything. Now hopefully we are moving into a period where those assumptions will be replaced with the more realistic ones that different actors have different information and learn in different ways. One approach to implementing these kinds of assumptions is available from control theory.

This paper discusses the learning procedures that are used in a variety of control theory methods. These methods begin with deterministic control with and without state variable and parameter updating. They also included two kinds of stochastic control: passive and active. With passive learning, stochastic control variables are chosen while considering the uncertainty in parameter estimates, but no attention is paid to the potential impact of today's control variables on future learning. By contrast, active learning control seeks a balance between reaching today's goals and gaining information that makes it easier to reach tomorrow's goals.

INTRODUCTION

We have just passed through a period in which the key assumptions in macroeconomic theory were that everyone knew everything. Now hopefully we are moving to a new period in which it is assumed that the various actors have different information about they economy; moreover, they learn but they do so in different ways.

Recently Abhay Pethe (1992) has suggested that we are now in a position to develop dynamic empirical macroeconomic models in which some actors learn in a sophisticated fashion by engaging in active learning with dual control techniques while other actors learn only incidentally as new observations arrive and are processed to form new estimates. One subset of this latter group considers the uncertainty in the economic system when they choose their actions for the next period. The other subset ignores the uncertainty in choosing a course of action for the next period. Finally, there is a fourth group that does not even bother to update their parameter estimates as additional observations are obtained.

While it is possible that one or more of these subgroups will be empty in any real economy, the starting assumption that different actors have different information, choose their actions in different ways, and learn in different ways seems a much more realistic and solid foundation for macroeconomics that the assumptions of the previous era.

D. A. Belsley (ed.), Computational Techniques for Econometrics and Economic Analysis, 75–87.

However, in the new period the analysis of macroeconomic systems will require different tools then those used in the previous era. While many results from the previous period could be obtained with analytical mathematics, the tools of the new era are much more likely to be computational. In anticipation of this, the current paper reviews the state of the art with regard to one set of tools that could serve well in the new era. These are the methods of control theory, which date back to the work of Simon (1956) and Theil (1957) as well as Aoki (1967), Livesey (1971), MacRae (1972), Prescott (1972), Pindyck (1973), Chow (1975) and Abel (1975). These methods are now enjoying a resurgence as attention turns once again to learning in economic systems.

Also the resurgence is being abetted by technical changes in computer hardware and software that have continued at a rapid pace in the last two decades. Control methods that were difficult to use twenty years ago on mainframe computers can now be used on ubiquitous desktop computers. Also super computers, some with parallel processing capabilities, are rapidly opening an era in which even active learning stochastic control methods can be used on economic models of substantial size.

It is in this context that this paper examines the current state of the art in numerical methods for control theory beginning with deterministic systems and passing through passive learning methods to end with active learning systems. The emphasis is not on the scope of the activity, since no attempt is made to be comprehensive. Rather the focus will be on areas where new developments in hardware and software offer us new opportunities. Also, some major problems that stand in our pathway will be highlighted. Deterministic problems will be discussed first, followed by passive learning and active learning problems.

1. DETERMINISTIC CONTROL

Since all uncertainty is ignored in solving deterministic control problems one is free to use either quadratic-linear or general nonlinear methods. Consider first quadratic-linear methods and then progress to the general nonlinear problems.

The deterministic quadratic-linear tracking problem is written as find

$$(u_k)_{k=0}^{N-1}$$

to minimize the cost functional

$$J = \tfrac{1}{2}[x_N - \tilde{x}_N]' W_N [x_N - \tilde{x}_N]$$

$$+ \tfrac{1}{2} \sum_{k=0}^{N-1} \{[x_k - \tilde{x}_k]' W [x_k - \tilde{x}_k] + [u_k - \tilde{u}_k]' \Lambda [u_k - \tilde{u}_k]\}, \tag{1}$$

where

x_k = state vector – an n vector,
$\tilde{x}_k$ = desired state vector – an n vector,

u_k = control vector – an m vector,
$\tilde{u}_k$ = desired control vector – an m vector,
W_N = symmetric state variable penalty matrix at terminal period, N,
W = symmetric state variable penalty matrix for periods 0 thru $N-1$,
Λ = symmetric control variable penalty matrix for periods 0 thru $N-1$,

subject to

$$x_{k+1} = Ax_k + Bu_k + Cz_k \ , \quad k = 0, \ldots, N-1 \ , \tag{2}$$

with x_0 given, where

A = state vector coefficient matrix $(n \times n)$,
B = control vector coefficient matrix $(n \times m)$,
C = exogenous vector coefficient matrix $(n \times \ell)$,
z_k = exogenous vector $(\ell \times 1)$ at time k.

In a macroeconomic setting the state variables are typically unemployment, inflation and the balance of payments, and the control variables are taxes, government spending and the money supply. Following Pindyck (1973) the problem is set up as tracking desired paths for both state and control variables as closely as possible.

The codes for solving this QLP (quadratic-linear) problem usually use Riccati equations and are very fast. They were originally coded in Fortran and later in Pascal and C. More recently they have been coded in metalanguages such as RATS and GAUSS. The RATS and GAUSS implementations have the advantage that the model can be estimated, simulated, and solved as an optimal control problem within the same framework.

However, one of the most effective methods of solving this class of models is the use of the GAMS language. GAMS is a modeling language that is set driven so that one can create a set of state variables and a set of control variable and then define mathematical relations over these sets. Thus it is not necessary to handcraft each equation but rather only types of equations that are defined over the sets. For an example of the solution of a quadratic linear macroeconometric model in the GAMS language see Parasuk (1989).

While the Riccati methods are confined to quadratic-linear models GAMS employs general nonlinear programming solvers such as MINOS to compute the solution to the model so the user can alter his or her problem from quadratic-linear to general nonlinear and continue using the same modeling system and solver. In addition the GAMS system is designed to be used with various solvers so, as technical progress is made in the solver software, the users gain the benefit of these changes without having to move the model from one code to another. An example is the recent addition of Drud's CONOPT (1992) software to the GAMS package. Thus the user can shift from using MINOS to using CONOPT by altering a single line in the GAMS problem representation. Since nonlinear programming codes have comparative advantages for different types of models, this ability to move easily between solvers could prove to be most beneficial.

Examples of the use of GAMS for general nonlinear macroeconometric control models are Fair's (1984) theoretical models for the household and the firm. These models have the advantage over analytic methods, viz Turnovsky (1977), that they can be extended to models with more than a few equations and can be solved for transitory as well as steady state solutions. The solutions contain the same kind of derivative sign information that was available with the analytical methods, viz when the money supply increases interest rates will fall. However, the numerical methods have the disadvantage that they yield results that hold only for the particular numerical parameters used. This loss though is mitigated by the fact that the analytical methods frequently encountered tradeoffs in which it was impossible to sign the outcomes. Thus even with its disadvantages, Fair's method provides a new and fresh approach to this part of macroeconomic theory.

Fair's original software implementation of this kind of modeling is relatively difficult to use. However, GAMS can be used to develop both household and firm models in an intuitive fashion so that the models can be easily altered. For an example of the Fair type of models in GAMS see Park (1992).

Deterministic models can be used with three types of learning – or the lack thereof. In the first type, the decision maker solves the deterministic model for a number of time periods and then uses this solution over the time horizon without solving the model again. Lets call this method "deterministic without update" or simply "deterministic".

In the second type, the decision maker solves the models for many time periods but only uses the policy values for the first time period. Then after the policy is applied and new values of the state variable emerge, he solves the problem again with these new state values as initial conditions. Once again, he uses only the first period policy and then repeats the process in each time period. This method can be called "deterministic with state update".

The third type is the same as the second except as new state variable observations come available in each period they are used to update parameter estimates in the system equations. This method can be called "deterministic with state and parameters update".

In summary "deterministic" decision makers are those who ignore the effects of uncertainty on the policy choice per se but they may engage in updating behavior with state variables and/or parameters. It seems likely that the largest group of decision makers fall into the second case. These people ignore uncertainty when making their decisions but they update the initial conditions for their dynamic problem each period as they move forward in time.

2. STOCHASTIC: PASSIVE LEARNING

Decision makers who use stochastic methods fall into two groups. Individuals in the first group use passive learning methods. Decision makers in this group consider the uncertainty in the system equation parameters while determining policies; however, no consideration is given to the effect of the decisions on future learning. In contrast, individuals who use active learning methods consider the possibility of perturbing the system in order to decrease parameter uncertainty in the future.

There are two sources of uncertainty in passive learning models: (1) additive error terms and (2) unknown parameters. There is also the possibility of state variable measurement error in passive learning models; however we delay the discussion of measurement error until the next section of this paper.

The most basic passive learning quadratic-linear tracking problem is written as find

$$(u_k)_{k=0}^{N-1}$$

to minimize the cost functional

$$J = E\{\tfrac{1}{2}[x_N - \tilde{x}_N]' \, W_N [x_N - \tilde{x}_N]$$

$$+\tfrac{1}{2} \sum_{k=0}^{N-1} [x_k - \tilde{x}_k]' \, W [x_k - \tilde{x}_k] + [u_k - \tilde{u}_k]' \, \Lambda [u_k - \tilde{u}_k]\} \, , \tag{3}$$

where

E = expectations operator,
x_k = state vector – an n vector,
$\tilde{x}_k$ = desired state vector – an n vector,
u_k = control vector – an m vector,
$\tilde{u}_k$ = desired control vector – an m vector,
W_N = symmetric state variable penalty matrix at terminal period, N,
W = symmetric state variable penalty matrix for periods 0 thru $N-1$,
Λ = symmetric control variable penalty matrix forperiods 0 thru $N-1$,

subject to

$$x_{k+1} = A(\theta_k)\, x_k + B(\theta_k)\, u_k + C(\theta_k)\, z_k + \xi_k \, , \quad k = 0, \ldots, N-1 \, , \tag{4}$$

with: x_0 given, where

A = state vector coefficient matrix $(n \times n)$,
B = control vector coefficient matrix $(n \times m)$,
C = exogenous vector coefficient matrix $(n \times \ell)$,
z_k = exogenous vector $(\ell \times 1)$ at time k,

and

$$\xi_k \sim N(0, Q) \, ,$$
$$\theta_0 \sim N(\hat{\theta}_0, \Sigma_{0|0}^{\theta\theta}) \, ,$$

where

ξ_k = normally distributed disturbance with zero mean and known covariance Q,
$\theta_k = s$ vector of unknown coefficients in A, B and C with initial estimates $\hat{\theta}_0$ and covariance $\Sigma_{0|0}^{\theta\theta}$ – both known,
$\Sigma_{0|0}^{\theta\theta}$ = known covariance matrix $(s \times s)$ for initial period parameter estimates,
Q = known covariance matrix $(n \times n)$ for system disturbances, ξ_k.

In this method, the covariance of the parameters of the systems equations, $\Sigma^{\theta\theta}$, plays a major role in the choice of controls. The policy makers avoid controls that add to the uncertainty in the system by choosing controls that are associated with parameters with low uncertainty or choosing combinations of controls that are associated with parameters that have negative covariances. Thus there is a motivation to hold a "portfolio" of controls that have relatively low uncertainty.

As is shown in Kendrick (1981, Ch. 6), passive learning controls can be computed with a variant of the Riccati method that is computationally very efficient. However, the calculations that involve the covariance of the parameters make this method somewhat less efficient than deterministic methods. Thus, the loss in moving from deterministic to passive learning stochastic methods is not computational efficiency so much as restriction on model specification. In deterministic methods one can easily move from quadratic-linear to general nonlinear specifications. However, stochastic control methods in the algorithms used in this paper are restricted to linear systems equations and normal distributions.

The reason for this restriction is that one needs to be able to map the uncertainty in one period into the next period with dynamic equations. It is a desirable property of such systems that the form of the distributions remain unchanged from one period to the next. For example, linear relationships can be used to map normal distributions in one period into normal distributions in the next period. In contrast, a quadratic relationship would map a normal distribution in one period into a chi square distribution in the next period and a Wishart distribution in the third period.

Restricting systems equations, and therefore econometric models, to linear equations is a high price to pay for being able to do stochastic control. Hopefully this restriction will soon be broken by advances in numerical methods. One promising approach to nonlinear models is Matulka and Neck (1992).

Passive learning models were formerly solved on mainframe computers. However, the personal computers in use today are fast enough to permit solution of these models on the desktop. For example, the DUAL code of Amman and Kendrick (1991) has recently been made available on IBM PC's and compatibles. This code has both passive and active learning capabilities, but for the time being it is expected that most usage on personal computers will be in the passive mode. This will change shortly with the widespread use of faster CISC and RISC microprocessors as is discussed below.

In summary, with passive learning stochastic control the choice of control variables in each period is affected by the covariance of the parameters of the system equation. Also, as with deterministic control there is updating of the parameter estimates and of the state variables in each period. We do not define separate names for the different updating behaviors because it seems sensible that any decision maker who is sophisticated enough to consider the covariance of the parameters in choosing his controls will also be sophisticated enough to update both parameter and state variable estimates in each time period.

There is also a good possibility that passive learning stochastic control methods can be applied to some game theory situations. Hatheway (1992) has developed a deterministic dynamic game model for the U.S. and Japanese economies using GAUSS. In Appendix B of his dissertation he outlines a method for extending his methodology to passive learning stochastic control models.

3. STOCHASTIC: ACTIVE LEARNING

Next we consider the actor who considers the effects of the choice of control variables in the current period on the future covariance of the parameters. This actor is sophisticated enough to realize that perturbations to the system today will yield improved parameter estimates that enable him to control the economic system better in the future. He is also sophisticated enough to know that if the elements in the covariance matrix are small that there will be little payoff to active learning efforts. Moreover he knows that even if the elements in the covariance matrix are small it may be worthwhile to attempt to learn if the additive system noises are large.

The model for this actor may be written as a general quadratic linear tracking problem which is to choose the control path

$$(u_k)_{k=0}^{N-1}$$

to minimize the cost functional

$$J = E\{\tfrac{1}{2}[x_N - \tilde{x}_N]' \, W_N [x_N - \tilde{x}_N]$$

$$+\tfrac{1}{2} \sum_{k=0}^{N-1} [x_k - \tilde{x}_k]' \, W [x_k - \tilde{x}_k] + [u_k - \tilde{u}_k]' \, \Lambda [u_k - \tilde{u}_k]\} \, , \tag{5}$$

where

E = expectations operator,
x_k = state vector – an n vector,
$\tilde{x}_k$ = desired state vector – an n vector,
u_k = control vector – an m vector,
$\tilde{u}_k$ = desired control vector – an m vector,
W_N = symmetric state variable penalty matrix at terminal period, N,
W = symmetric state variable penalty matrix forperiods 0 thru $N - 1$,
Λ = symmetric control variable penalty matrix forperiods 0 thru $N - 1$,

subject to

$$x_{k+1} = A(\theta_k) \, x_k + B(\theta_k) \, u_k + C(\theta_k) \, z_k + \xi_i \, , \quad k = 0, \ldots, N-1 \, , \tag{6}$$

with: x_0 given,
where

A = state vector coefficient matrix $(n \times n)$,
B = control vector coefficient matrix $(n \times m)$,
C = exogenous vector coefficient matrix $(n \times \ell)$,
z_k = exogenous vector $(\ell \times 1)$ at time k,
ξ_k = additive system error term.

The measurement relations are

$$y_k = H(\theta_k)\,x_k + \zeta_k \tag{7}$$

and the first order Markov process

$$\theta_{k+1} = D\theta_k + \eta_k\ , \tag{8}$$

where

y_k = measurement vector – an r-vector,
H = measurement coefficient matrix $(r \times n)$,
ζ_k = measurement error term – r-vector for each period,
D = known Markov process matrix $(s \times s)$,
η_k = time-varying parameter error term – s-vector for each period,

where the vectors ξ_k, ζ_k, η_k, x_0, $\Sigma_{j|j}^{\theta\theta}$ are assumed to be mutually independent, normally distributed random vectors with known means and covariances (positive semi-definite):

initial period state:	$x_0 = N(\hat{x}_0, \Sigma_{0\|0}^{xx})$,
initial parameters:	$\theta_0 = N(\hat{\theta}_0, \Sigma_{0\|0}^{\theta\theta})$,
system noise:	$\xi_k = N(0, Q)$,
measurement noise:	$\zeta_k = N(0, R)$,
Markov process noise:	$\eta_k = N(0, G)$,

and where

$\Sigma_{0|0}^{xx}$ = known covariance matrix $(n \times n)$ for initial period state variables,
$\Sigma_{0|0}^{\theta\theta}$ = known covariance matrix $(s \times s)$ for initial period parameter estimates,
Q = known covariance matrix $(n \times n)$ for system disturbances, ξ_k,
R = known covariance matrix $(r \times r)$ for measurement disturbances, ζ_k,
G = known covariance matrix $(s \times s)$ for Markov disturbances, η_k.

Measurement error is also included in this model. Thus the state variables are not observed directly but rather through a noisy process. Of course as the sizes of the measurement errors decrease the gain to active learning efforts will increase.

The presence of measurement error in models with distributed lags also raises the following issue: Normally some data are collected in each time period, and flash estimates are issued before the full data set has been collected and processed. Thus the most recent state estimate will be the noisiest while state variables from several periods ago will have less noise associated with them. So there is a premium on using data from several periods ago in the feedback rule. However, to control a system well one wants to use the most recent state variables. This tradeoff between recent states with noisy measurements and lagged states with less noisy measurement has not yet been studied numerically. However, the computer code to facilitate such work is already available.

The problem setup here is general enough to include not only measurement errors but also to permit inclusion of time varying parameters. This level of sophistication has not yet been programmed into our numerical codes, but the mathematical derivations and separate program development have been done by Tucci (1989). When time varying parameters are present, the parameter covariance elements are likely to be larger, so there will be more gain from active learning efforts. On the other hand parameters learned today will be changing in the future; therefore there is less potential gain from learning. This tradeoff has not yet been studied numerically.

Active learning stochastic control can be done with the DUAL code mentioned above. This program has recently been modified and versions developed for supercomputers and workstations as well as mainframes. We have versions running on Cray and IBM supercomputers, IBM mainframes and SUN and IBM workstations. In addition a version for IBM PC's and compatibles has recently been developed. We have discovered that we can solve small active learning problems even on IBM AT computers with 80286 chips and substantially larger models on IBM PS/2 computers with 80386 chips. Thus we are confident that the 486 chips and beyond will have the capability to solve active learning stochastic control problems with a number of states and controls.

Also we have found that it is possible to do large numbers of Monte Carlo runs on very small models using SUN and IBM workstations. So far these experiments have shown that actors who are sophisticated enough to employ active learning techniques will not necessarily perform better on average than actors who use passive learning stochastic control methods or even in some cases deterministic methods, cf. Amman and Kendrick (1994). However, we are treating these results with some caution because of the possibility that nonconvexities in the cost-to-go can affect them.

More than ten years ago Kendrick (1978) and Norman, Norman, and Palash (1979) first encountered nonconvexities in active learning stochastic control problems. However, these results were obtained with computer codes of such complexity that it was uncertain whether or not the nonconvexities were fundamental or not. Also, the codes and computers of that time were not fast enough to permit detailed studies of the problem. However, recently Mizrach (1991) has cast new light on this problem by providing detailed derivations for the single-state, single-control problem of MacRae (1972). He found that the nonconvexity was not a passing phenomenon but rather was fundamental to active learning problems solved with the Tse and Bar-Shalom(1973) algorithm.

Amman and Kendrick (1992) then followed Mizrach's work by using numerical work to confirm his results and focusing on the cause of the nonconvexities as the initial covariance of the unknown parameter. As an aid to understanding this result consider the MacRae model that is stated below.

The MacRae model was chosen for this work because it is the simplest possible adaptive control problem. If nonconvexities occur in this problem then one can expect that they will also appear in more complex models. The MacRae model is

find (u_0, u_1) to minimize

$$J = E\{\tfrac{1}{2}w_2 x_2^2 + \tfrac{1}{2}\sum_{k=0}^{1}(w_k x_k^2 + l_k u_k^2)\} \tag{9}$$

subject to

$$x_{k+1} = a x_k + b u_k + c + e_k\ , \quad \text{for } k = 0, 1, \tag{10}$$

$$x_0 = 0\ . \tag{11}$$

The parameter values used by MacRae are

$$a = .7\ , \ b = -.5\ , \ c = 3.5\ , \ s_e^2 = q = .2$$

$$w_k = 1\ \forall k\ , \ l_k = 1\ \forall k\ , \ s_b^2 = .5\ , \ s_a^2 = s_c^2 = 0\ .$$

Also the desired paths in (9) are implicitly set to zero so

$$\tilde{x}_k = 0\ , \ \tilde{u}_k = 0 \quad \forall k\ .$$

This problem has been solved using the dual control algorithm of Tse and Bar-Shalom as described in detail in Ch. 11 of Kendrick (1981). At period k of an N period model with $N-k$ periods to go to total cost-to-go can be written as

$$J_{N-k} = J_{D,N-k} + J_{C,N-k} + J_{P,N-k}\ , \tag{12}$$

where the D, C, and P subscripts represent the deterministic, cautionary, and probing components, respectively. The deterministic term includes all of the nonstochastic elements. The cautionary term is a function of $\Sigma_{k+1|k}$, i.e. of the uncertainty in the next period before a new control can be applied. The probing term can be written as

$$J_{P,N-k} = \tfrac{1}{2}\, tr \sum_{j=k+1}^{N-1}(\mathcal{R}_j\, \Sigma_{j|j}^{bb})\ , \tag{13}$$

where $\Re$ is a Riccati-like term and $\Sigma_{j|j}^{bb}$ is the covariance matrix of the unknown parameters in period j after updating with data through period j. Notice that the probing term is a function of the parameter covariance matrix for all periods from the current to the terminal period.

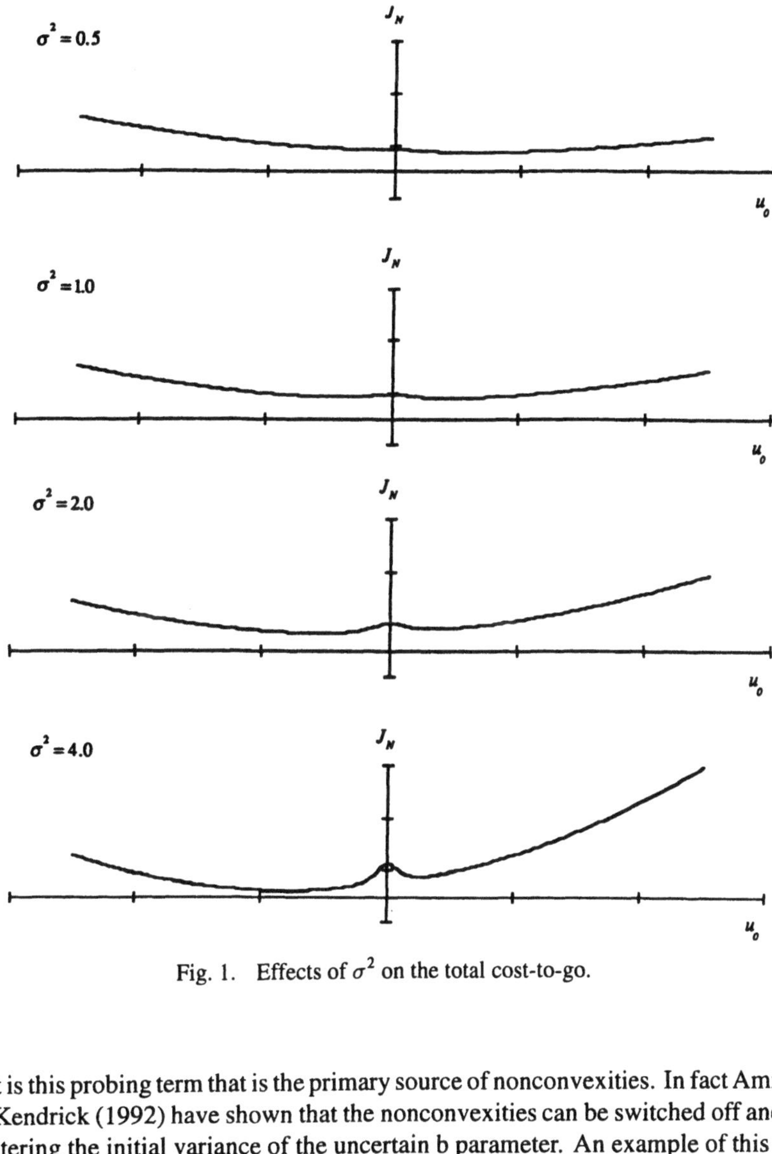

Fig. 1. Effects of σ^2 on the total cost-to-go.

It is this probing term that is the primary source of nonconvexities. In fact Amman and Kendrick (1992) have shown that the nonconvexities can be switched off and on by altering the initial variance of the uncertain b parameter. An example of this sort is shown in Figure 1. With the setting of $\sigma^2 = .5$ at the top of the figure the cost-to-go function remains a convex function of the initial period control, u_0. However, as σ^2 increases the nonconvexity appears and causes two local optima for the problem.

In addition to the nonconvexities from the probing term, Amman and Kendrick also found that there are combinations of parameter values which, in conjunction with large values of σ^2, will result in nonconvexities also arising in the cautionary term.

So the bad news is that the nonconvexities appear to be fundamental to active learning stochastic control problems. However, the good news is that there may be some regularities about these nonconvexities that can be exploited to design efficient solution algorithms. Also, even if brute force grid search methods must be employed, computer speeds are increasing so rapidly that models that exhibit nonconvexities can be solved.

Finally, there is some prospect that the parameter values in empirical economic models will be such that the nonconvexities occur only rarely. It will take some time and effort to establish this fact, but one can be hopeful that this will occur.

4. CONCLUSIONS

Economists do not need to use the unrealistic assumption that all economic actors know everything. Rather there are tools at hand that will allow us to portray different actors as having different information and able to learn as time passes. Moreover, there are available algorithms and computer codes for modeling different kinds of learning behavior in different actors. Some actors may be so sophisticated as to use active learning methods in which they probe the system in order to improve parameter estimates over time. Other actors may be sophisticated enough to consider the covariance of parameters in choosing their control variables but learn only passively with the arrival of new information. Other actors may be so unsophisticated that they do not even update parameter estimates when new observations arrive.

Since computer speeds have increased greatly in recent years we can now model all these kinds of behaviors using code that operates on supercomputers, workstations and even personal computers.

However, the most sophisticated methods that involve active learning can give rise to nonconvexities in the cost-to-go, so caution must be exercised until we can learn more about when these nonconvexities arise and how to solve active learning problems when they do occur.

REFERENCES

Abel, Andrew (1975), "A Comparison of Three Control Algorithms to the Monetarist-Fiscalist Debate," *Annals of Economic and Social Measurement*, Vol. 4, No. 2, pp. 239–252, Spring.

Amman, Hans M. and David A. Kendrick (1991), "A User's Guide for DUAL, A Program for Quadratic-Linear Stochastic Control Problems, Version 3.0", Technical Paper T90–94, Center for Economic Research, The University of Texas, Austin, Texas 78712.

Amman, Hans M. and David A. Kendrick (1992), "Nonconvexities in Stochastic Control Models", Paper 92–91, Center for Economic Research, The University of Texas, Austin, Texas, 78712.

Amman, Hans M. and David A. Kendrick (1994), "Active Learning – Monte Carlo Results," forthcoming in 1994 in Vol. 18 of the Journal of Economic Dynamics and Control.

Aoki, Masanao (1967), *Optimization of Stochastic Systems*, Academic Press, New York.

Chow, Gregory (1975), *Analysis and Control of Dynamic Systems*, John Wiley and Sons, Inc., New York.

Drud, Arne (1992), "CONOPT – A Large Scale GRG Code," forthcoming in the *ORSA Journal on Computing*.

Fair, Ray (1984), *Specification, Estimation and Analysis of Macroeconometric Models*, Harvard University Press, Cambridge, Mass. 02138.

Hatheway, Lawrence (1992), *Modeling International Economic Interdependence: An Application of Feedback Nash Dynamic Games*, Ph.D. Dissertation, Department of Economics, The University of Texas, Austin, Texas 78712.

Kendrick, David A. (1978), "Non-convexities from Probing an Adaptive Control Problem," *Journal of Economic Letters*, Vol. 1, pp. 347–351.

Kendrick, David A. (1981), *Stochastic Control for Economic Models*, McGraw-Hill Book Company, New York.

Livesey, David A. (1971), "Optimizing Short-Term Economic Policy," *Economic Journal*, Vol. 81, pp. 525–546.

MacRae, Elizabeth Chase (1972), "Linear Decision with Experimentation," *Annals of Economic and Social Measurement*, Vol. 1, No. 4, October, pp. 437–448.

Matulka, Josef and Reinhard Neck (1992), "A New Algorithm for Optimum Stochastic Control on Nonlinear Economic Models," forthcoming in the *European Journal of Operations Research.*

Mizrach, Bruce (1991), "Non-Convexities in an Stochastic Control Problem with Learning," *Journal of Economic Dynamics and Control*, Vol. 15, No. 3, pp. 515–538.

Norman, A., M. Norman and C. Palash (1979), "Multiple Relative Maxima in Optimal Macroeconomic Policy: An Illustration", *Southern Economic Journal*, 46, 274–279.

Parasuk, Chartchai (1989), *Application of Optimal Control Techniques in Calculating Equilibrium Exchange Rates*, Ph.D. Dissertation, Department of Economics, The University of Texas, Austin, Texas 78712.

Park, Jin-Seok (1992), *A Macroeconomic Model of Monopoly: A Theoretical Simulation Approach and Optimal Control Applications*, Ph.D. dissertation in progress, Department of Economics, University of Texas, Austin, Texas 78712.

Pethe, Abhay (1992), "Using Stochastic Control in Economics: Some Issues", Working Paper 92–5, Center for Economic Research, The University of Texas, Austin, Texas, 78712.

Pindyck, Robert S. (1973), *Optimal Planning for Economic Stabilization*, North Holland Publishing Co., Amsterdam.

Prescott, E. C. (1972), "The Multi-period Control Problem under Uncertainty," *Econometrica*, Vol. 40, pp. 1043–1058.

Simon, H. A. (1956), "Dynamic Programming under Uncertainty with a Quadratic Criterion Function," *Econometrica*, Vol. 24, pp. 74–81, January.

Theil, H. (1957), "A Note on Certainty Equivalence in Dynamic Planning," *Econometrica*, Vol. 25, pp. 346–349, April.

Tse, Edison and Yaakov Bar-Shalom (1973), "An Actively Adaptive Control for Linear Systems with Random Parameters," *IEEE Transactions on Automatic Control*, Vol. AC–17, pp. 38–52, February.

Tucci, Marco (1989), *Time Varying Parameters in Adaptive Control*, Center for Economic Research, The University of Texas, Austin, Texas 78712.

Turnovsky, Stephen J. (1973), "Optimal Stabilization Policies for Deterministic and Stochastic Linear Systems", *Review of Economic Studies*, Vol. 40.

Turnovsky, Stephen J. (1977), *Macroeconomic Analysis and Stabilization Policy*, Cambridge University Press, London.

ALFRED LORN NORMAN

Computability, Complexity and Economics

ABSTRACT. Herbert Simon advocates that economists should study procedural rationality instead of substantive rationality. One approach for studying procedural rationality is to consider algorithmic representations of procedures, which can then be studied using the concepts of computability and complexity. For some time, game theorists have considered the issue of computability and have employed automata to study bounded rationality. Outside game theory very little research has been performed. Very simple examples of the traditional economic optimization models can require transfinite computations. The impact of procedural rationality on economics depends on the computational resources available to economic agents.

1. INTRODUCTION

H. Simon (1976) suggests that the proper study of rationality in economics is procedural rationality. Simon believes that procedural rationality should encompass the cognitive process in searching for solutions to problems. This study should be performed using computational mathematics, which he defines as the analysis of the relative efficiencies of different computational processes for solving problems of various kinds. "The search for computational efficiency is a search for procedural rationality, ..." In this paper, problem-solving processes are formalized as algorithms for solving economic problems. Placed in an algorithmic format, procedural rationality can be studied using the theory of computability and complexity developed by mathematicians and computer scientists.

In Section 2 the concepts of computability and complexity are presented. The traditional format of computability is for finite representations. One example is finite sequences from a finite alphabet. Another is the study of functions $f : N^n \rightarrow N^k, n \geq 0, k > 0$, where N is the natural numbers $0, 1, 2, \ldots$. While this model is appropriate for studying finite state game theory, it is not applicable to most traditional single agent optimization problems such as the theory of the firm or the consumer defined as optimization problems over the reals. To study the complexity of such problems, the information-based complexity concept of Traub, Wasilkowski and Woźniakowski (1988) is recommended. This approach encompasses both finite representable combinatorial complexity as well as optimization over the reals. An important question in complexity theory is whether a problem is tractable, that is can be computed with polynomial resources.

One application of complexity theory is determining the computational cost of achieving accuracy in algorithms used in numerical analysis such as integration. Economists should perform such analyses for algorithms used in optimization models

D. A. Belsley (ed.), Computational Techniques for Econometrics and Economic Analysis, 89–108.

and econometrics. A start in this direction has been made by Norman and Jung (1977), Norman (1981, 1994) and Rustem and Velupillai (1987) in the area of linear quadratic control. In this paper, we focus on the relationship between computability and complexity and economic theory with special emphasis on bounded rationality. In this paper we focus on computational complexity and do not consider dynamic complexity arising from chaotic behavior even though Spear (1989) demonstrates that the two concepts are related.

In section 3 the literature concerning computability, complexity and bounded rationality in finite action game theory is considered. This literature dates back at least to Rabin's (1957) demonstration of the existence of a noncomputable strategy. More recently Binmore (1990) and Canning (1992) have considered the impact of restricting players to computable algorithms. Since Aumann's (1981) suggestion, game theorists have modeled bounded rationality by replacing players with automata. A brief survey of this literature is presented. For automata theory there are two types of complexity: the computational complexity of computing the best-response automaton and the strategic complexity of implementing the strategy. Overall, game theory contains many problems currently considered intractable.

Outside of game theory very little research in economics has been done on computability and complexity and their relationship to bounded rationality. The literature concerning the theory of the firm and the theory of the consumer is considered in Section 4. Norman (1994) demonstrates that very simple models of the firm can require transfinite computations to determine profit maximization. Also, such transfinite problems cannot be ignored by appealing to concepts such as ϵ-rationality, because the computational complexity of ϵ-optimization can be exponential, that is, intractable. Beja (1989) and Rustem and Velupillai (1989) demonstrate a fatal flaw in the traditional choice model. Norman (1992) proposes a new discrete-mathematics consumer model for choice with technological change.

In Section 5 we briefly consider two miscellaneous, unrelated articles. The first is Spear's (1989a) use of computability theory to characterize the identification of a rational expectation equilibrium. The second is Norman's (1987) use of computational complexity to characterize alternative mechanisms to clear the Ostray-Starr (1974) household exchange problem.

Section 6 forecasts the impact of computability and complexity on economics. If bounded rationality is interpreted as optimization with a computational resource restriction, the impact on economic theory depends on whether the restriction is computability, tractability or linearity.

Finally, the reader is warned that because symbol usage generally follows the references, some symbols are used for several purposes in the paper.

2. COMPUTABILITY AND COMPLEXITY

There are several approaches to the theory of computability that include recursive functions, Turing machines, algorithms, and rewrite systems. Because these alternatives are equivalent up to a coding (transformation), the choice selected should be that most accessible to the reader. While mathematicians, and hence economic theorists, generally prefer the recursive function approach, the readers of *Computational Economics* are likely to prefer an algorithmic approach that is intuitively obvious to economists with some computer programming experience.

Let us consider the algorithmic approach to computability of Sommerhalder and van Westrhenen (1988), which analyzes the properties of simple-algorithmic-language, SAL(**N**), programs. A SAL(**N**) program is a mathematical entity defined as a quadruple (n, k, p, P), where P is a sequence of SAL statements, and the variables occurring in the sequence P belong to $x_1, \ldots, x_p \in N^p$. Of these p variables, $n \geq 0$ are input variables and $k > 0$ are output variables. There are two types of SAL statements:

1. Assignment statements: (Note: we will use $\leftarrow$, not :=, for assignment)

$x_i \leftarrow 0$

$x_i \leftarrow x_j$

$x_i \leftarrow x_j + 1$ (Successor)

$x_i \leftarrow x_j \diamond 1$ (if $x_j = 0$ then $x_i \leftarrow 0$ else $x_i \leftarrow x_j - 1$) (Predecessor)

2. While statement:

while $x_i \neq 0$ **do** S **od** where S is a sequence of SAL statements.

The set F(SAL(**N**)) contains all functions $f{:}N^n \to N^k$, $n \geq 0$, $k > 0$ for which there exists a SAL(**N**) program (n, k, p, P) which, given an input$(x_1, \ldots, x_n)$, computes $(x_1, \ldots, x_k) = f(x_1, \ldots, x_n)$ as output in a finite number of steps. If the program computes an output for at least one input, the function is a partial recursive function, and if the program computes an output for every input, the function is a total recursive function. The set F(SAL(**N**) is equivalent to the set of recursive functions. (See Sommerhalder and van Westrhenen, 1988).

In SAL programs, arithmetic operations such as +, −, ×, and ÷ must be constructed as macros. For example the addition macro $x_i \leftarrow$ ADD$(x_i + x_j)$ can be constructed as

while $x_j \neq 0$ **do**
$x_i = x_i + 1$
$x_j = x_j \diamond 1$
od

Adding the standard arithmetic operations as statements in SAL would decrease the number of statements required to compute a function; nevertheless, if a function was not computable without arithmetic statements, it would not become computable with arithmetic statements. Computability addresses the issue of what can be computed in a finite number of statements, not the number of statements. For simplicity, it is desirable to keep the instruction set of SAL to the minimum.

A focus of computability theory is *decidability* - - that is whether a predicate can be determined in a finite number of steps. One of these predicates is the halting problem for SAL(**N**) programs: Does computation on a given input x, induced by a given program P, terminate? This problem is not solvable by algorithmic methods; that is we can not construct a universal program to answer this question. A famous example is Hilbert's tenth problem: Can we construct an algorithm to determine whether a given polynomial equation with integer coefficients has an integer solution? Matijasevitš (1971) determined that such an algorithm cannot be constructed. Also, computability is closely related to Gödel's work on the limitations of constructive mathematics. Indeed, a frequently used concept in computability proofs is the Gödel index scheme.

An economic example that clarifies the concept of decidability is the use of order and rank conditions to determine if a simultaneous structural econometric model is identified. Since these conditions can be checked by a program running in polynomial time, the issue is decidable. However, determining the precise values of the unknown, identified parameters is not decidable, because (besides being real numbers) these are asymptotic limits that cannot be determined with a finite number of observations and calculations.

Complexity theory addresses the issue of categorizing the difficulties of alternative problems that can be solved in a finite number of steps. For an overview of formal complexity theory see Hartmanis (1989). With the exception of Blum's (1967) axiomatic approach, theorists concerned with the properties of complexity classes, such as whether *P* is a proper subset of *NP*, have traditionally used the Turing machine as their model of computation since it provides a common frame of reference for considering such relationships.

However, because the Turing model is very tedious to apply to specific finite-representable combinatorial problems and this model is not applicable without modification for most real-number numerical analysis problems, complexity practitioners have constructed many computer models appropriate for the study of particular problems and algorithms. For an overview of such models and complexity applications, see Aho, Hopcropft and Ullman (1974). For example, a straight line program model is generally used in matrix multiplication analysis, and a decision tree, for which each node represents a comparison, is generally used in sorting analysis. Also, applied complexity analysis frequently uses a cost function to reflect the cost of performing the operation central to the algorithm in question. Nevertheless, because most common models of computation are polynomial related, and the asymptotic definitions of complexity only count the most frequently occurring operation, the various models are closely related.

Because most economic optimization problems are defined over the reals – not the natural numbers – we need a notion of computational complexity general enough to deal with both types of formulations. Two major branches of computational complexity are *information-based complexity* and *combinatorial complexity*. The former deals with the difficulty of approximating solutions to problems where information is partial, noisy, and costly. The later deals with problems that can be solved exactly in a finite number of computations and for which information is complete, exact, and costless. Most combinatorial problems can be represented using natural numbers or finite sequences from a finite alphabet. In this paper we need a computational model that can be used to study both information-based and combinatorial complexity. For this purpose we employ a slightly modified version of the information-based computational model of Traub, Wasilkowski and Woźniakowski (1988). Here all arithmetic and combinatorial operations are assumed to be performed with infinite precision.

Let the economic problem set be designated

$$F = \{f(x_1, \ldots, x_n) : (x_1, \ldots, x_n) \in I \in \Re^n\}, \tag{1}$$

where I is the input set. The solution operation is $S_T : F \to G$, where G is a normed linear space. In cases where G is not a normed linear space, there is a generalized solution operator that need not be discussed in this paper. Associated with each problem element is a solution element $S(f)$. Let $U(f)$ be the computed approximation to $S(f)$ with absolute error measured by $\mid S(f) - U(f) \mid$. We shall say that $U(f)$ is an *ϵ-approximation* iff $\mid S(f) - U(f) \mid \leq \epsilon$.

To compute these ϵ-approximations we may need information about f. We gather knowledge about f through the use of information operations $\Gamma : F \to H$. For each problem element $f \in F$, we compute a number of information operations, which can either be adaptive or nonadaptive. Associated with the set of information operations $\Gamma = \{\gamma_1, \ldots, \gamma_L\}$ is a cost vector $C'_\Gamma = \{c_{\gamma_1}, \ldots, c_{\gamma_L}\}$. In numerical analysis, an example of an information operation would be the cost necessary to obtain the value of a function at a point in an integration procedure based on function evaluations. In economics, information operations could be used to represent the cost of acquiring data in the marketplace. The knowledge of f obtained through information operators is represented as $N(f)$.

Given the information obtained from the information operators, the ϵ-approximations are computed using a specified set of combinatory operations, $\Omega = \{\omega_1, \ldots, \omega_K\}$. Associated with these combinatory operations is a cost vector $C'_\Omega = \{c_{\omega_1}, \ldots, c_{\omega_K}\}$. The operations to be included in Ω constitutes an important component of the model of computation. For SAL(**N**), these operations are the "assignment" and "while" statements. For the study of the computational complexity of numerical analysis problems, Ω consists of arithmetic operations, comparison of real numbers, and the evaluation of certain elementary functions. For economic problems, we will introduce additional operators. Some of these information and combinatory operators will be considered *oracles*, that is black boxes that can perform an operation with a specified cost. We do not consider how the black box performs the operation.

For each $f \in F$, we desire to compute a ϵ-approximation $U(f)$ of the true solution $S(f)$, where $\epsilon = 0$ corresponds to an exact solution. From knowing $N(f)$, the approximation $U(f)$ is computed by a mapping ϕ that corresponds to an *algorithm*, where $U(f) = \phi(N(f))$, with

$$\phi : N(f) \to G, \tag{2}$$

and the goal is to compute $\phi(N(f))$ at minimal cost. If no information is required, $\phi(N(f))$ reduces to $\phi(f)$. This very generalized conception of an algorithm is called an *idealized* algorithm. Much complexity analysis is performed by restricting idealized algorithms to realizable algorithms that are based on a particular computer model, such as a Turing machine, or on computational considerations, such as the class of algorithms that are linear functions of the input.

The cost of information gathering and computing $\phi(N(f))$, which will be denoted by $\varphi_1(\phi, N(f))$,is

$$\varphi(U, f) = C'_\Gamma \Psi(\Gamma) + C'_\Omega \Psi(\Omega), \tag{3}$$

where U stands for a pair consisting in information N and algorithm ϕ. $\Psi(\cdot)$ is a vector whose ith element is the number of operations performed on the ith element of Γ or Ω, as designated. This cost function is closely related to the time needed to perform the computation. To determine the total time, the cost vectors would be replaced with the time needed to perform the associated operations.

In this paper we concern ourselves only with the worst-case setting of complexity. Here the error and cost of approximation are defined over all problem elements as follows:

$$e(U) = \sup_{f \in F} \mid S(f) - U(f) \mid, \tag{4}$$

$$\varphi(U) = \sup_{f \in F} \varphi(U, f). \tag{5}$$

Another important complexity concept is the average complexity. Formulating the average complexity requires knowing the distribution of the occurrence of the elements of f in F. Since such knowledge is not available for most economic problems, we instead consider the range of performance over F. The cost function as defined is a measure of the transactions cost of decision making. One consideration is to compare the absolute costs of alternatives. Another important consideration is how these costs grow with increasing problem size, which we shall designate by the generic parameter T. For SAL(**N**) problems, $T = n$, the number of inputs. In considering the asymptotic cost function an important question is whether the growth of these costs is no more than a polynomial in T. Such problems are considered *tractable*. Since as T increases, the cost is progressively dominated by highest power of the polynomial, the definitions for asymptotic complexity assign problems and algorithms to equivalence classes based on this highest power.

We wish to compare $\varphi(U(T))$ with a nonnegative $Z = Z(T)$, which in applications will frequently be T, T^2, T^3 and so on.

Definition 1. $\varphi(U(T))$ is of upper order [lower order] Z, written $O(Z)[o(Z)]$, if there exist $k, m > 0$ such that $\varphi(U(T)) \leq [\geq]\, mZ(T)$ for all $T > k$.

Definition 1 requires a slight modification to handle the rate of growth measured in terms of achieving greater accuracy, that is $1/\epsilon \to \infty$. This definition can now be employed to characterize the computational complexity of the two optimization problems by applying the definition of upper and lower order to the cost functions of Φ which is the class of all algorithms that use information operator N.

Definition 2. F has ϵ-computational complexity Z if there exists an ϵ-approximate algorithm $U(f) \in \Phi$ such that $\varphi(T)$ is $O(Z)$ and, for all ϵ-approximate algorithms $U \in \Phi$, $\varphi(T)$ is $o(Z)$.

Like definition 1, definition 2 requires a slight modification to handle the cost of achieving greater accuracy as measured by $1/\epsilon$. To say that F has 0-computational complexity T^0 means that F can be computed exactly (ϵ=0) in a fixed number of computations independent of the length of the time horizon T. Definition 1 divides algorithms into equivalence classes. For example, an algorithm which can compute F in six operations is equivalent to one that can compute F in eight. For algorithms whose cost functions are polynomial in T, the equivalence classes are defined by the highest power of T. For asymptotic analysis, the cost of the operation that is performed with the highest power of T can be assigned a value of 1 and all the other information and combinatory operations can be assigned a value of 0. Thus in analyzing sorting algorithms, only the number of comparisons is considered. If the concern in analyzing problems is to determine which problems are tractable, the problem formulation is reasonably robust to the selection of elements of Ω, because most standard computational models are polynomial related.

In this paper we consider only one complexity class, which contains problems currently considered intractable, namely the nondeterministic polynomial *NP*. While *P* and *NP* are usually defined relative to deterministic and nondeterministic Turing machines, let us consider defining them relative to SAL(**N**) and ND-SAL(**N**) to avoid introducing a new model.

To discuss the *NP* class we have to add a statement to SAL to create the nondeterministic simple algorithmic language, ND-SAL(**N**). The new statement is

3. Either statement: **either** sequence, S_i **or** sequence, S_j **od**

The intent of the **either** statement is that one of the two sequences S_i or S_j will be executed. However, which one is left undetermined. In a SAL program the computational sequence is a straight line. In an ND-SAL(**N**) program, one path in a tree is executed.

To illustrate the operation of an ND-SAL(**N**) program consider the partition problem: Given a set Q of natural numbers, does there exist a subset $J \subseteq Q$ such that

$$\sum_{x \in J} x = \sum_{x \in (Q-J)} x? \tag{6}$$

For simplicity, consider the special case where Q consists of just three numbers. We introduce a SAL macro for addition called ADD. The critical steps in an ND-SAL(**N**) program would be three statements (i = 1,2,3):

either $x_4 \leftarrow$ ADD($x_4 + x_i$) **or** $x_5 \leftarrow$ ADD($x_5 + x_i$).

The program would terminate only if x_4 equals x_5. After these three statements have been executed there are eight possibilities:

Case	x_4	x_5
1	$x_1 + x_2 + x_3$	0
2	$x_1 + x_2$	x_3
3	$x_1 + x_3$	x_2
4	$x_2 + x_3$	x_1
5	x_1	$x_2 + x_3$
6	x_2	$x_1 + x_3$
7	x_3	$x_2 + x_3$
8	0	$x_1 + x_2 + x_3$

If any of the eight possibilities discovers a partition, the ND-SAL(N) program terminates successfully. In the equivalent SAL(N) program, at least four of the eight possibilities must be considered.

Having briefly introduced ND-SAL(N), let us define *P* and *NP*. Defining unit time and cost for executing a statement, polynomial cost and polynomial time are equivalent. A problem F is a member of the polynomial class *P* [nondeterministic polynomial class *NP*] if there exists a SAL(N) [ND-SAL(N)] program that can solve each member f of F as a polynomial function of the number of inputs, n.

In terms of computability, ND-SAL(N) is no more powerful than SAL(N) because any function that can be computed by ND-SAL(N) can be computed by SAL(N). Nevertheless, programs in ND-SAL(N) can be construed as countably parallel in comparison to the equivalent program in SAL(N). Thus a ND-SAL(N) program that solves F in polynomial time could have a separate polynomial path for each f. The equivalent SAL(N) program could consider all these paths in exponential time. One of the most famous open questions in computer science is whether there exist problems in *NP* which are not members of *P*.

A well-known group of problems in *NP*, which are assumed not to be members of *P*, are known as *NP*-complete. To show that a new problem is *NP*-complete requires two steps. First, a solution that runs in polynomial time must be verified. Second, one of the existing *NP*-complete problems must be polynomial transformable into the new problem. There are numerous *NP* complete problems including many operation research problems, such as the traveling salesman problem, and many graph problems, such as the Hamilton circuit problem. These problems currently require exponential time or cost in SAL(N). For an introduction to *NP* complete problems see Papadimitriou and Steiglitz (1982).

3. BOUNDED RATIONALITY AND GAME THEORY

Game theory is the only field of economics that has generated a literature concerning computability, complexity, and bounded rationality. This is not totally surprising since finite action game theory is one of the few economic subjects fitting the traditional computational models for problems represented by either natural numbers or finite sequences from a finite alphabet. The first topic we consider is the impact of the concept of computability on game theory. Next we consider an example of an *NP*-complete problem in game theory and finally, we consider finite automata as a form of bounded rationality.

The knowledge that there exist games with noncomputable optimal strategies has been known at least since Rabin's (1957) paper. We now present a simple number theoretic example due to Jones (1982).

Example 1. An arithmetical Game of Length 5

There are two players 1 and 2 who take turns assigning nonnegative integer values to the variables of a polynomial:

player 1 picks x_1
player 2 picks x_2
player 1 picks x_3
player 2 picks x_4
player 1 picks x_5

Player 1 wins if and only if $x_1^2 + x_2^2 + 2x_1x_2 - x_3x_4 - 2x_3 - 2x_5 - 3 = 0.$ *Otherwise player 2 wins.* In any arithmetical game, either player 1 has a winning strategy or player 2 does. But Jones provides a specific example where neither player 1 nor player 2 has a computable winning strategy. This example is related to the undecidability of Hilbert's 10th problem.

Recent work in computability in game theory has investigated the impact of imposing a restriction of computability on player strategies. We assume each player is replaced by an algorithm that generates a strategy choice given a complete description of the games, including the other player's algorithm. Such an algorithm is *complete* if it produces a strategy choice in every situation. It is also *rational* if it generates the optimal response to the other player's choices. Binmore (1990) demonstrates that computable, complete, and rational algorithms do not exist.

Canning (1992) investigates relaxing completeness to obtain algorithms that are rational and complete on a limited domain. Let $H \subseteq G$, the set of games with finite strategies, and let $B \subseteq A$, the set of effectively computable game theories. (H, B) is solvable if there exists a strategy in A that is complete relative to (H, B) and is the best choice whenever the opponent plays. Canning demonstrates that (H, A) is solvable if and only if $H \subseteq D$, where D is the set of games with dominant strategies for each player. Also, (G, K) is solvable if K is the set of algorithms that always stop. These results define the limits of rational, computable games. To explore these limits Canning develops concepts such as a strict Nash strategy, a best reply to all best replies to itself, and a rational algorithm that plays the best response if the opponent reaches a decision. He encounters a basic problem that the set of rational algorithms of A is too small to include any algorithm that acts rationally and is complete against every algorithm in the set of rational algorithms of A.

In addition to the investigation of computability in game theory, researchers have used complexity theory in game theory investigations. Prasad and Kelly (1990) provides examples of *NP*-completeness in determining properties of weighted majority voting games. Such a game consists of n individuals making up the set $N = \{1, 2, \ldots, n\}$ with an associated vector of weights $W = (w_1, w_2, \ldots, w_n)$. A weighted majority voting game is one in which, for some fixed q, coalition $S \subseteq N$ is winning just when $\sum_{j \in S} w_j \geq q$.

Given nonnegative integer weights and a positive integer q, the question of determining the existence of a subset $S \subseteq N$ such that $\sum_{j \in S} w_j = q$ is known to be *NP*-complete. Prasad and Kelly use this problem to examine the complexity of determining various power measures of $i \in N$. i is pivotal in subset $S \subseteq N - \{i\}$ if $\sum_{j \in S} w_j < q$ and $w_i + \sum_{j \in S} w_j \geq q$. Most power measures are functions of the number of distinct subsets for which i is pivotal. Prasad and Kelly show that determining whether the number of pivots is greater than r is an *NP*-complete problem. They also show the standard political power indices, such as the Absolute Banzhaf, Banzhaf-Colemen and Shapley-Subik, are all *NP*-complete problems.

Imposing a computability constraint on a game is not likely to create controversy among economists. First, the constraint appears obvious; and second, it is robust to the choice of computational model. The best way to impose tractability on games is more controversial. Since suggested by Aumann (1981), game theorists have considered automatons as players in games in order to study bounded rationality. Kalai (1990) provides an excellent survey of this literature. Here we need only a short summary based on the bimatrix representation of the stage game for a repeated prisoner's dilemma

		Player 2's Actions	
		c	d
Player 1's Actions	c	(3, 3)	(0, 4)
	d	(4, 0)	(1, 1)

The payoff matrix for this game is symmetric. Both players' 1 and 2 actions are c, cooperate, and d, defect. The first entry in each element of the payoff matrix represents the payoff to player 1 and the second the payoff to player 2.

This game has one Nash equilibrium (d, d). While both players would be better off cooperating (c, c), this action combination is not stable because both players could improve their position by switching actions. In the repeated prisoner's dilemma, the problem is to determine the circumstances under which the two players would cooperate to achieve a higher payoff. Intuitively it would seem likely that they would have incentives to cooperate. Let us consider this problem. Let $a^t \in \{(c, c), (c, d), (d, c), (d, d)\}$ be the action combination selected by players 1 and 2 in period t. A history h of length $l(h) = k$ is $\{a^j, a^{j+1}, \ldots, a^{j+k}\}$, and H^T is a set of all histories of length strictly less than T. A strategy for player i in period t is $f_i^t : H^{t-1} \rightarrow (c, d)$; that is, a strategy provides a rule for action given all possible histories. One method of calculating the payoffs in a repeated game is the average payoff. Let $f^1 = (f_1(h^0), f_2(h^0))$, which means the strategy for the first stage. Then, recursively for $t = 2,3,\ldots,T$: $f^t = f(f^1, f^2, \ldots, f^{t-1})$. Let $P(f^t)$ be the payoff to the two players in period t when they use strategy combination f^t. Then the average payoff to the two players is $\bar{P}(f) = (1/T)\sum_{t=1}^{T} P(f^t)$.

Let us now describe two very simple strategies for the repeated prisoner's dilemma. Since the game is symmetric, we only need describe the strategies for player 1. These two are

a. Constant defect: $f_1(h) \to d$ for all histories h.
b. Tit-for-tat: $f_1(h^0) \to c$ and $f_1(h) \to h_2^l$. (That is, initially cooperate and afterwards execute the last action taken by player 2.)

As the game is repeated, the set H^t of all possible histories , which is also the domain of the strategy, increases exponentially. Nevertheless, for the average payoff, T-period, repeated prisoner's dilemma game, the only Nash equilibrium is the action combination (d, d) each period.

Kalai's approach for studying bounded rationality in repeated games is full automation, where both players are replaced by automata. An automaton is a triple $((M, m^0), B, T)$, where M is the set of states of the automaton. The behavior function $B : M \to (c, d)$ prescribes an action to player 1 at every state on the automaton. The transition function $T : M \times A \to M$ transits the automaton to a new state from an old one as a function of the action combinations of both players. The automata for the two strategies listed above are presented in the following table.

Strategies of player 1	States	B	$\binom{c}{c}$	$\binom{c}{d}$	$\binom{d}{c}$	$\binom{d}{d}$
Constant defect	$m^0 = D$	d	D	D	D	D
Tit-for-Tat	$m^0 = C$	c	C	D	C	D
	$m^0 = D$	d	C	D	C	D

Neyman (1985) maintains that the two-person, repeated prisoner dilemma played by automata can result in a cooperative Nash equilibrium. Let $\bar{P}^T_{m_1,m_2}$ represent T repetitions with the average payoff criterion, where each player i chooses an automaton of size not exceeding m_i. Neyman asserts that, if $2 \leq m_1,\ m_2 \leq T - 1$, then there is a Nash equilibrium pair of automata of $\bar{P}^T_{m_1,m_2}$ that prescribes cooperation throughout $\bar{P}^T$. The occurs because restricting the size of the automata prevents the usual backward induction. Zemel (1985) introduces small talk into the finitely repeated prisoner's dilemma as an alternative approach to explaining cooperation.

Next let us consider Pen-Porath's (1986) result concerning the advantage of having a bigger automaton in an infinitely repeated two-person zero-sum game, $\bar{Z}^{\infty}_{m_1,m_2}$. Since zero-sum games have a value in mixed strategy, every player can guarantee his/her pure strategy Z-maxmin value with an automaton of size one. Ben-Porath's result concerning the advantage of being bigger is that for every given positive integer m_1, there is a positive integer m_2, and an automaton A_2 of size m_2, such that for every automaton A_1 of size m_1, player 1's payoff is no more than the pure strategy Z-maxmin value of player 1.

Rubinstein (1986) and Abreu and Rubinstein (1988) have investigated the choice of automata when the number of states is costly. Also, games with finite actions have the desirable property that all equilibrium payoffs can be well approximated by equilibria of bounded complexity. This idea is pursued in the papers of Kalai and Stanford (1988) and Ben-Porath and Peleg (1987).

In addition to characterizing the behavior of automata, game theorists have also investigated the computational complexity of computing the best-response automaton under various conditions. Gilboa (1988) considers the problem of computing the best-response automaton, A_1, for player 1 in a repeated game G with n players and $n-1$ finite automata, $(A_2, \ldots, A_n)$, for the remaining players in G. He demonstrates that the computational complexity of both problem (1) – determining whether a particular A_1 is a best-response automaton – and problem (2) – finding a best-response automaton A_1 – is polynomial. If the number of players is unrestricted, problem (1) is *NP*-complete and problem (2) is not polynomial. Ben-Porath (1990) demonstrates that for a repeated two person game where player 2 plays a mixed automaton strategy with a finite support, problem (1) is a *NP*-complete problem and problem (2) does not have a polynomial solution. Papadimitriou (1992) considers the relationship between the computational complexity of determining a best-response strategy and the strategic complexity in a repeated prisoner's dilemma. If an upper bound is placed on the number of states of the best-response automaton, the problem is a *NP*-complete problem; whereas, if no bound is imposed, the problem is polynomial.

Finally, game theorists are in the process of developing a complexity measure for implementing an automaton. Kalai-Stanford (1988) define the *complexity of a strategy* to be its size (the number of states of the smallest automaton prescribing it). In general the amount of information needed for playing a strategy equals the complexity of the strategy; that is, the complexity of a strategy, f, equals the number of equivalence classes of histories it induces. Banks and Sundaram (1990) propose an alternative strategic complexity concept that includes a measure of the need to monitor the opponent's action. Lipman and Srivastava (1990) propose a strategic complexity measure based on the details of the history required by the strategy. They are interested in the frequency with which perturbations in history change the induced strategy. Papadimitriou's (1992) result indicates that achieving a specified Kalai-Stanford strategic complexity increases the computational complexity of computing the best response automaton.

4. THE FIRM AND THE CONSUMER

The original calculus-based models of profit and utility maximization are defined over the reals – for example, the positive orthant of $\Re^n$. Consequently, the traditional computability and complexity arguments based on either the natural numbers or finite representations from a finite alphabet are not applicable.

In order to demonstrate just how simple a noncomputable optimization problem can be, we consider the problem presented in Norman (1994), which employs the information-based complexity model. A monopolist has a linear production process, faces a linear inverse demand function, and has a profit function for $t = 1,2,\ldots,T$:

$$p_t = a - dq_t,$$

$$\pi_t = p_t q_t - c(q_t), \tag{8}$$

$$q_t = \beta x_t + \zeta_t,$$

where a and d are known, q_t is the tth observation of net output, x_t is the tth level of the production process, β is the unknown scalar parameter, and ζ_t is the tth unobserved disturbance term. The ζ_t are iid normal with mean zero and known variance one. Since the complexity results are invariant to defining the cost function as a zero, linear, or quadratic function, the cost function is defined as $c(q_t) = 0$ to simplify the notation.

Given a normal prior on β at time $t = 1$, the prior information on β at time t is a normal distribution $N(m_t, h_t)$, where m_t is the mean updated by $h_t = h_{t-1} + x_{t-1}^2$and h_t is the precision updated by $m_t = (m_{t-1}h_{t-1} + q_{t-1}x_{t-1})/h_t$. For this paper let us consider two cases:

1. The agent knows β precisely. He or she has either been given precise knowledge of β or has observed (1) a countable number of times so that his or her prior on β has asymptotically converged to $N(\beta, \infty)$.

2. The agent's prior information on β is represented by $N(m_1, h_1)$, where h_1 has a very small positive value.

The monopolist is interested in maximizing his expected discounted profit over a finite time horizon:

$$J_T = \sup_{x^T} E[\sum_{t=1}^{T} \tau^{t-1} p_t(x_t) q_t(x_t) \mid q^{t-1}, x^{t-1}], \tag{9}$$

where τ is the discount factor, q^{t-1} is $(q_1, q_2, \ldots, q_{t-1})$ and x^{t-1} is $(x_1, x_2, \ldots, x_{t-1})$. q^{t-1} and x^{t-1} represent the fact that the decision maker anticipates complete information that is observed exactly and without delay.

First consider the optimization problem where β is a known parameter. The optimal x_t can be exactly determined as a function of the parameters of $f \in F$ without recourse to the information operator as

$$x_t^* = \frac{a}{2d\beta}. \tag{10}$$

The (ϵ=0)-computational complexity of this problem is T^0, polynomial zero, because the control that can be computed in 3 operations needs to be computed only once for the entire time horizon.

Now let us illustrate the computational difficulty with case 2, the simplest non-trivial example having a time horizon of only two periods, $T = 2$. The value function in the first period is

$$J_1(q_1) = \frac{a^2((m_1 h_1 + q_1 x_1)/h_1)^2}{4d([(m_1 h_1 + q_1 x_1)/(h_1 + x_1^2)]^2 + (h_1 + x_1^2)^{-1})} - d. \tag{11}$$

While the expectation of $J_1(q_1)$ has the form

$$E\left[\frac{Q_1(q_1)}{Q_2(q_1)}\right] - d, \tag{12}$$

where $Q_1(q_1)$ and $Q_2(q_1)$ are quadratic forms in the normal variable q_1. This expectation cannot be carried out explicitly to give an analytic closed expression. This implies the 0-complexity of this problem with an unknown parameter is transfinite.

Norman (1993) uses these two cases to provide a Bayesian explanation of Knight's concepts of risk and uncertainty. Risk is where the parameters and distributions of the decision problem are known, and uncertainty is where at least one parameter or distribution is not known. The conjecture is that, for nonlinear problems, the ϵ-computational complexity of an uncertainty problem always lies in a equal or higher computational class than that for the equivalent risk problem.

The reader might have an illusion that transfinite problems are an oddity in economics. The author asserts that the opposite is likely to be the case. Readers who are not familiar with computational complexity, but who have some knowledge of numerical analysis, should realize that all those problems for which the traditional numerical analysis focused on asymptotic convergence of alternative algorithms are transfinite computational problems. The author asserts that most of the standard calculus optimization problems in the theory of the consumer and the firm are transfinite. Only special cases, such as quadratic problems, are computable. Also, expressions that are defined by infinite series are frequently not computable. Another example is traditional asymptotic convergence theory of econometric estimates.

The reader might assume that the problem can be circumvented by appealing to ϵ-rational arguments; that is, by using ϵ-approximations which can be computed in a finite number of computations. If the constraint is that these approximations be tractable in the sense of being polynomial costs with respect to the growth parameters of the problem, using ϵ-approximations is not always possible.

Consider the discrete-time, stationary, infinite horizon discounted stochastic control problem requiring computation of a fixed point J^* of the nonlinear operator T (acting on a space of function on the set $S \in \Re^n$) defined by Bellman's equation

$$(TJ)(x) = \inf_{u \in U} [g(x,u) + \alpha \int_S J(u) P(y|x,u) dy], \qquad \forall x \in S. \tag{13}$$

Here, $U \subset \Re^m$ is the control space, $g(x,u)$ is the cost incurred if the current state is x and control u is applied, $\alpha \in (0,1)$ is a discount factor, and $P(y|x,u)$ is a stochastic kernel that specifies the probability distribution of the next state y when the current state is x and control u is applied. Then J^* is interpreted as the value of the expected discounted cost, starting from state x, providing that the control actions are chosen optimally. A variation of this model has been considered by economists [for example Easley and Keifer (1989)] investigating parameter estimation in an estimation and control context. By treating unknown parameters as augmented states, the simple monopoly model presented in this section could be generalized to n states and m controls over an infinite horizon.

Chow and Tsitsiklis (1989) show that the computational complexity of this model is $o(1/[k(\alpha)\epsilon]^{2n+m})$. This means that, for a given accuracy, the computation cost is exponential in increasing model size (number of states and controls). Thus, to assume ϵ-rationality in general is to assume economic agents have exponential computing power.

Another area of traditional economic theory for which ideas of computability and complexity have been considered is consumer theory. A choice function is a system of pairwise preferences that to be preference-compatible must select from every set of feasible alternatives those maximal elements that are undominated. Beja (1989) proves "that for the class of choice functions whose domain includes all finite sets and some infinite set(s), a characterization (by axioms of rational choice) of compatibility with preferences which are not necessarily transitive and complete must include some *infinite complexity axiom*, i.e. an axiom that posits simultaneous consistency across infinite collections of decisions." Such a condition is obviously not decidable in any model of computation that must consider these collections sequentially.

Velupillai and Rustem (1990) consider the choice problem from the perspective of a nondeterministic Turing machine. They present a nondeterministic Turing machine with a Gödel-numbered sequence of finite sets of alternatives and inquire whether the Turing machine for each pair (x, y) in each sequence can determine whether x is at least as good as y. They demonstrate that there is no finite procedure to answer this question; that is, the issue is not decidable. Velupillai and Rustem's results imply the standard choice model has fundamental computational problems without even considering an infinite complexity axiom needed for consistency of preferences.

Norman (1992) considers a simple model of a consumer choosing an item from a finite set of close substitutes $B_t = \{b_{1t}, b_{2t}, \ldots, b_{nt}\}$ for $t = 1, 2, \ldots, T$, either once or repeated at regular intervals. The consumer problem in period t, S_t, is

$$\text{Find a } b_{it}^* \in B_{ft} \text{ such that for all } b_{jt} \in B_{ft} \text{ is } b_{it}^* \succeq b_{jt}, \tag{14}$$

where p_{it} is the price of the ith item in the tth period, I_t is the income in the tth period, and $b_{it} \in B_{ft}$ if $b_{it} \in B_t$ and $p_{it} \le I_t$.

Because of the high rate of technological change in the marketplace and the usually long time interval between purchases, the consumer of durable goods generally faces a new set of alternatives possessing new technological attributes. We assume that the consumer searches for his preferred item by ranking his alternatives. This ranking operation is costly, because it requires real resources in the form of mental effort, time, and travel expenses. Given the rapid rate of technological change, we assume that the consumer's preferences are not given *a priori* but are determined, to the extent they can be done so efficiently, in the consumer's search for the preferred item.

We model the ranking of two items as a binary operation, $R(b_{it}, b_{jt})$, which the consumer must execute to determine his preferences between two items, b_{it} and b_{jt}. This operation is modeled as a primitive operation with positive costs, and no attempt is made to model the human neural network. We assume that the cost, c, of comparing items is invariant to the two items being compared. The reflexive binary ranking operation $R(b_{it}, b_{jt})$ is assumed to have the following cost: Given any two unranked b_{it} and $b_{jt} \in B$, $C(R(b_{it}, b_{jt}))$, the cost of executing $R(b_{it}, b_{jt})$ is c. If b_{it} and b_{jt} have been ranked, the cost of remembering $R(b_{it}, b_{jt})$ is 0. Also the consumer could rank alternatives if he or she choose. However, given the cost, this might not be optimal. In addition, the consumer expends resources to determine which items in his or her consumption set are budget feasible: For any $b_{it} \in B$, the cost of performing $F(b_{it})$ is k.

The consumer's search to find an optimal consumption bundle depends on market organization. The type of organization considered is a consumer selecting a new TV from a wall of TVs presented in a electronics discount store. Consequently, the consumer's search can be conceptualized as one through an unordered sequence to find a preferred item satisfying a budget constraint; the consumer's search can be modeled as an algorithm. Organized in this fashion, characterizing an efficient search is equivalent to determining the combinatorial computational complexity of the choice problem.

The computational complexity of finding the preferred item in a one-time choice problem is n. An efficient algorithm, then, is a variation of finding the largest number in a sequence. Thus, in a one-time choice problem, it is never efficient to develop a complete preference ordering, which is a variation of sorting a file and has a computational complexity of $n \ln n$. Consequently, if ranking alternatives is expensive, a procedural rational consumer facing technological change would never determine a complete preference ordering, a fundamental assumption of a substantive rational consumer.

5. TWO PAPERS

In this section we consider two separate, unrelated papers. First, Spear (1989) demonstrates how the imposition of computability on a rational expectations equilibrium, REE, with incomplete information implies that such equilibria are not identifiable. Second, Norman (1987) demonstrates how complexity theory can be used to create a theory of money. These papers are related only in that they provide some insights into the range of topics in economics to which the concepts of computability and complexity might be applied.

Spear considers a two-period overlapping-generations model. To use finite representation computability theory, he assumes the economy has a countable number of states, S. The set Φ consists of total recursion functions on S, where total means that the associated SAL(**N**) programs stop for all states $s \in S$. The economy maps admissible forecasts ϕ_0 into temporary equilibria, T.E. price functions $\phi_1 \in \Phi$. This mapping, $\hat{g} : \Phi \to \Phi$, which, given the assumptions is $g : N \to N$, is assumed total recursive and has a fixed point.

Spear considers the problem of determining the circumstances under which agents can identify the rational expectations equilibrium, REE. For the problem under consideration, identification means the ability to construct an algorithm that can decide in a finite (not asymptotic) number of steps which function among a class of recursive functions has generated an observed sequence of ordered pairs of numbers of the form $(j, f[j])$. The two basic results for complete information are (1) if the T.E. price function is primitive recursive, agents can identify it; however, if $\phi_{g[i]}$ is not primitive recursive, identification may not be possible, and (2) if the function g is primitive recursive, it can be identified in the limit. Primitive recursive functions are those that can be computed by SAL(**N**) programs that do not employ **while** statements. (Sequences of assignment statements can be executed a specified number of times with **times** statements.)

With incomplete information the basic result is: *There is no effective procedure for determining when a given model-consistent updating scheme yields a REE, unless* R_g *is empty.*

In the second paper, Norman (1987) constructs a theory of money based on the complexity of barter exchange.

The monetary model employed is the Ostray-Starr (1974) household exchange problem: Let W and Z be $n \times H$ matrices representing the initial endowments and excess demands of the H households with columns representing households and rows representing goods. The entries of W are non-negative. A positive entry for Z indicates an excess demand, a negative entry excess supply. Given an n-vector price p whose elements are all positive, the system (p, Z, W) satisfies for $i = 1,2,\ldots,n$ and $j = 1,2,\ldots,H$ the following restrictions:

$$p'Z = 0,$$

$$\sum_{j=1}^{H} Z_{ij} = 0, \tag{7}$$

$$Z_{ij} \geq -W_{ij}.$$

These conditions state that the value of each household's excess demands equals the value of its excess supplies, and the excess supply of any good cannot exceed its respective endowment. In addition, aggregate excess demand equals aggregate excess supply.

In this model the general equilibrium auctioneer has generated a set of equilibrium prices, and the task remains to find a set of trades that clear the resulting household excess demands. In a manner analogous to the creation of the auctioneer, a broker is created to arrange a clearing sequence, a set of trades that will reduce all household excess demands to zero. The difficulty of the broker's task depends on the conditions imposed on each trade. For all exchange mechanisms considered, all trades considered must satisfy the condition that the value of the goods received by a household must equal the value of goods sent without credit. If no other conditions are imposed on the exchange mechanism, the broker can simply exchange all excess demands simultaneously. The computational complexity of the resulting "command exchange" mechanism is nH.

Because bilateral barter will not clear the household exchange model in general, multiparty barter in the form of chains is considered. In a chain, household j_1 receives good i_1 and sends good i_2. Household j_2 receives good i_2 and sends good i_3. Household j_m receives good i_m and sends good i_1. The value of the goods being traded, y, is equal in all cases. The computational complexity of the multiparty barter exchange mechanism is the minimum of (n^2H, nH^2). Introducing money reduces the complexity of the exchange mechanism to nH.

6. CONCLUDING REMARKS

To be consistent with Lipman (1991), we define bounded rationality as optimization with restricted computational resources where the optimizing procedure is specified as an algorithm. The impact of this definition of bounded rationality on economics depends on the computational resources available to the optimizing economic agent. Imposing a constraint of computability on economic agents is not likely to be contested by many economists. This restriction will have some impact on economic theory. As was pointed out, Rustem and Velupillai (1990) demonstrate that choice theory will have to be reformulated.

Many economists would accept a definition of bounded rationality as tractability, that is polynomial computational resources. Currently *NP*-complete problems require exponential resources in deterministic models of computation. Assuming a polynomial solution to these problems does not exist, such a definition would have a major impact on economics because numerous *NP*-complete problems exist in current economic theory. For example, the use of automata in game theory would have to be refined. Also, in many cases of optimization over the reals, the concept of ϵ-rationality could not be maintained.

Most humans do not sort large files or perform conventional matrix multiplication of any size without machine assistance. This might suggest that an appropriate bound on computational complexity might be a low order polynomial. The author asserts that economic agents unaided by machines are restricted to algorithms which are at most linear in the growth parameters. The impact of such a restriction on economic theory would be massive. While the author believes such a bound is appropriate, his opinion may not be shared by many economists.

REFERENCES

1. Abreu, D. and Rubinstein, A. 1988, "The structure of Nash equilibrium in repeated games with finite automata", *Econometrica*, vol 56, No. 6.
2. Aho, A. V., J. E. Hopcroft and J. D. Ullman, 1974, *The design and analysis of computer algorithms* (Addison-Wesley: Reading).
3. Aumann, R. J., 1981, "Survey of repeated games", in *Essays in Game theory and Mathematical Economics in Honor of Oskar Morgenstern* (Bibiographische Institut: Mannheim).
4. Banks, J. S. and R. K. Sundaram, 1990, "Repeated games, finite automata, and complexity", *Games and Economic Behavior*, vol 2, pp. 97–119.

5. Beja, A. 1989, "Finite and infinite complexity in axioms of rational choice or Sen's characterization of preference-compatibility cannot be improved", *Journal of Economic Theory*, vol 49, pp. 339–346.
6. Ben-Porath, E. 1986, "Repeated games with finite automata.", IMSSS, Stanford University (manuscript).
7. Ben-Porath, E. and Peleg, B. 1987, "On the Folk theorem and finite automata", The Hebrew University (discussion paper).
8. Ben-Porath, E., 1990, "The complexity of computing a best response automaton in repeated games with mixed strategies", *Games and Economic Behavior*, vol 2, pp. 1–12.
9. Binmore, K. 1990, *Essays on the Foundations of Game Theory*, (Basil Blackwell, Oxford).
10. Blum M., 1967, "A machine independent theory of the complexity of recursive functions", *J. ACM*, vol 14, pp. 3322–336.
11. Canning, D. 1992, "Rationality, Computability, and Nash Equilibrium", *Econometrica*, Vol 60, No 4, pp. 877–888.
12. Chow, Chee-Seng and John N. Tsitsiklis, 1989, "The Complexity of Dynamic Programming", *Journal of Complexity*, 5, 466–488.
13. Easley, David and N. M. Keifer, 1988, "Controlling a stochastic process with unknown parameters", *Econometrica*, Vol 56 No. 5, 1045–1064.
14. Jones, J. P., 1982, "Some Undecidable Determined Games", *International Journal of Game Theory*, vol. II, Issue 2, pp. 63–70.
15. Gilboa, Itzhak, 1988, "The complexity of computing best-response automata in repeated games", *Journal of Economic Theory*, vol 45, pp. 342–352.
16. Hartmanis, Juris, 1989, "Overview of Computational Complexity Theory in Hartmanis", J (ed) *Computational Complexity Theory* (American Mathematical Society: Providence).
17. Kalai, E., 1990, "Bounded Rationality and Strategic Complexity in Repeated Games", in Ichiishi, T, A. Neyman, and Y. Tuaman, (eds), *Game Theory and Applications*, (Academic Publishers, San Diego).
18. Kalai, E. and W. Stanford, 1988, "Finite rationality and interpersonal complexity in repeated games", *Econometrica*, vol 56, 2, pp. 397–410.
19. Lipman, B. L. and S. Srivastava, 1990, "Informational requirements and strategic complexity in repeated games", *Games and Economic Behavior*, vol 2, pp. 273–290.
20. Lipman, B. L. 1991, "How to decide how to decide how to...: Modeling limited rationality", *Econometrica*, vol 59, No. 4, pp. 1105–1125.
21. Matijasevis̃, J. V., 1971, "On recursive unsolvability of Hilbert's tenth problem", Proceedings of the Fourth International Congress on Logic, Methodology and Philosophy of Science, Bucharest, Amsterdam 1973, pp. 89–110.
22. Neyman, A., 1985, "Bounded complexity justifies cooperation in the finitely repeated prisoner's dilemma", *Economics Letters*, Vol 19, pp. 227–229.
23. Norman, A., 1981, "On the control of structural models", *Journal of Econometrics*, Vol 15, pp. 13.24.
24. Norman, Alfred L., 1987, "A Theory of Monetary Exchange", *Review of Economic Studies*, 54, 499–517.
25. Norman, Alfred L., 1992, "On the complexity of consumer choice, Department of Economics", The University of Texas at Austin, (manuscript) Presented at the 1992 Society of Economics and Control Summer Conference, Montreal.
26. Norman, Alfred L., 1994, "On the Complexity of Linear Quadratic Control", *European Journal of Operations Research*, 73, 1–12.
27. Norman, Alfred L., 1994, "Risk, Uncertainty and Complexity", *Journal of Economic Dynamics and Control*, 18, 231–249.
28. Norman, Alfred L. and Woo S. Jung, 1977, "Linear Quadratic Control Theory For Models With Long Lags", *Econometrica*, 45, no.4, 905–917.
29. Ostroy, J. and R. Starr, 1974, "Money and the Decentralization of Exchange", *Econometrica*, vol 42, pp. 1093–1113.

30. Papadmimitriou, C. H., 1992, "On players with a bounded number of states", *Games and Economic Behavior*, Vol 4, pp. 122–131.
31. Papadminitriou, C. H. and K. Steiglitz, 1982, *Combinatorial Optimization: Algorithms and Complexity*, (Prentice-Hall: Englewood Cliffs).
32. Prasad K. and J. S. Kelly, 1990, "NP-Completeness of some problems concerning voting games", *International Journal of Game Theory*, Vol 19, pp. 1–9.
33. Rabin, M. O., 1957, "Effective computability of winning strategies", M. Dresher et al. (eds), *Contributions to the Theory of Games*, Annals of Mathematical Studies, Vol 39, pp. 147–157.
34. Rubinstein, A. 1986, "Finite automata play the repeated prisoner's dilemma", *Journal of Economic Theory*, vol 39, pp. 83–96.
35. Rustem, B. and K. Velupillai, 1987, "Objective Functions and the complexity of policy design", *Journal of Economic Dynamics and Control*, vol 11, pp. 185–192.
36. Rustem, B and K. Velupillai, 1990, "Rationality, computability, and complexity", *Journal of Economic Dynamics and Control*, vol 14, pp. 419–432.
37. Simon, H. A., 1976, "Form substantive to procedural rationality", S. Latsis (ed), *Method and Appraisal in Economics*, (Cambridge University Press, Cambridge).
38. Sommerhalder, R. and S. van Westrhenen, 1988 *The theory of Computability: Programs, Machines, Effectiveness and Feasibility*, (Addison Wesley: Workingham).
39. Spear, S. E., 1989a, "When are small frictions negligible?", in Barnett, W., J. Geweke, and K. Shell (eds), *Economic complexity: Chaos, sunspots, bubbles, and nonlinearity*, (Cambridge University Press, Cambridge).
40. Spear, S. E., 1989, "Learning Rational Expectations under computability constraints", *Econometrica*, Vol 57, No. 4, pp. 889–910.
41. Traub, J.F., G. W. Wasilkowski and H. Woźniakowski, 1988, *Information Based Complexity*, (Academic Press, Inc., Boston).
42. Zemel, E., 1985, "Small talk and cooperation: A note on bounded rationality", *Journal of Economic Theory*, vol 49, No. 1, pp. 1–9.

BERÇ RUSTEM

Robust Min-max Decisions with Rival Models

ABSTRACT. In the presence of rival models of the same system, an optimal policy can be computed to take account of all the models. A min-max, worst-case design, problem is an extreme case of the ordinary pooling of the models for policy optimization. It is shown that, due to its noninferiority, the min-max strategy corresponds to the robust policy. If such a robust policy happens to have too high a political cost to be implemented, an alternative pooling can be formulated using the robust pooling as a guide.

An algorithm is described for solving the constrained min-max problem. This consists of a sequential quadratic programming subproblem, a stepsize strategy based on a differentiable penalty function and an adaptive rule for updating the penalty parameter.

The global convergence and local convergence rate of the algorithm are established in Rustem (1992). In this paper, we discuss the numerical convergence properties of the algorithm and related issues such as the convergence of the stepsize to unity and the properties of the penalty parameter.

1. INTRODUCTION: THE POLICY OPTIMIZATION PROBLEM

Consider the policy optimization problem

$$\min \left\{ J(Y,U) \,\middle|\, F(Y,U) = 0 \right\}, \tag{1}$$

where Y and U are, respectively, the endogenous or output variables and policy instruments or controls of the system. J is the policy objective function and F is the model of the economy. In general, F is nonlinear with respect to Y and U. Problem (1) is essentially a static transcription of a dynamic optimization problem in discrete time, where

$$U = \begin{bmatrix} u_1 \\ \vdots \\ u_t \\ \vdots \\ u_T \end{bmatrix} \; ; \quad Y = \begin{bmatrix} y_1 \\ \vdots \\ y_t \\ \vdots \\ y_T \end{bmatrix} ,$$

with $u_t \in \mathbb{R}^n$ and $y_t \in \mathbb{R}^m$ denoting the control and endogenous variable vectors at time period t. The optimization covers the periods $t = 1, \ldots, T$. Thus, $Y \in \mathbb{R}^{m\times T}$, $U \in \mathbb{R}^{n\times T}$, $F : \mathbb{F} \subset \mathbb{R}^{n_x} \rightarrow \mathbb{R}^{T\times m}$ and $J : \mathbb{J} \subset \mathbb{R}^{n_x} \rightarrow \mathbb{R}^1$ $(n_x = T \times (m+n))$. The vector valued function F is essentially an econometric model comprising a

D. A. Belsley (ed.), Computational Techniques for Econometrics and Economic Analysis, 109–134.

system of nonlinear difference equations represented in static form for time periods $t = 1, \ldots, T$.

2. RIVAL MODELS OF THE SAME SYSTEM AND ROBUSTNESS OF MIN-MAX POLICIES

The formulation of the policy optimization problem (1) is, in practice, an over-simplification. Originating from rival economic theories, there exist rival models purporting to represent the same system. The problem of forecasting under similar circumstances has been approached through forecast pooling by Granger and Newbold (1977) and, more recently, by Makridakis and Winkler (1983) and Lawrence *et al.*, (1986). In the presence of rival models, the *policy maker may also wish to take account of all existing rival models in the design of optimal policy.* One strategy in such a situation is to adopt the worst case design problem

$$\min_{Y^1,\ldots Y^{m_{\text{mod}}},U} \max_i \left\{ J^i(Y^i,U) \;\middle|\; F^i(Y^i,U) = 0;\; i = 1,\ldots,m_{\text{mod}} \right\}, \tag{2}$$

where there are $i = 1, \ldots, m_{\text{mod}}$ rival models, with Y^i, F^i, respectively, denoting the dependent (or endogenous) variable vector and the equations of the ith model. This strategy is an extension of a suboptimal approach originally discussed in Chow (1979). Problem (2) seeks the optimal strategy corresponding to the most adverse circumstance due to choice of model. All rival models are assumed known. The solution of (2) clearly does not provide insurance against the eventuality that an unknown $(m_{\text{mod}} + 1)$st model happens to represent the economy; it is just a robust strategy against known competing "scenarios". A similar, less extreme formulation is also discussed below, utilizing the dual approach to (2).

The optimization procedure considered below does not distinguish between Y and U. We can thus define a general vector $x = \begin{bmatrix} Y \\ U \end{bmatrix}$ to rewrite the min-max problem as follows:

$$\min_x \max_i \left\{ J^i(x) \;\middle|\; F(x) = 0,\; i = 1,\ldots,m_{\text{mod}} \right\}, \tag{3}$$

where F subsumes all the models. The formulation above is slightly more general than the original min-max problem above. Other equivalent formulations are discussed in Rustem (1987, 1989, 1992).

Algorithms for solving (3) have been considered by a number of authors, including Charalambous and Conn (1978), Coleman (1978), Conn (1979), Demyanov and Malomezov (1974), Demyanov and Pevnyi (1972), Dutta and Vidyasagar (1977), Han (1978; 1981), Murray and Overton (1980). In the constrained case, discussed in some of these studies, global and local convergence rates have not been established (e.g. Coleman, 1978; Dutta and Vidyasagar, 1977). In this and the next section, a dual approach to (3), adopted originally by Medanic and Andjelic (1971, 1972) and Cohen (1981), is initially utilized. Subsequently, both dual and primal approaches are used to formulate a superlinearly convergent algorithm.

To introduce the basic terminology, let $x \in \mathbb{R}^{n_x}$ and let $F : \mathbb{R}^{n_x} \to \mathbb{R}^{m \times T}$ be twice continuously differentiable functions with

$$\hat{J} = [J^1, J^2, \ldots, J^{m_{\text{mod}}}]^T .$$

Let 1 be the m_{mod}-dimensional vector whose elements are all unity. We define the inner product of two vectors, y and w, of the same dimension as

$$\langle y, w \rangle \equiv y^T w .$$

Using the inner product, we define the subspace

$$\mathbb{R}_+^{m_{\text{mod}}} = \left\{ \alpha \in \mathbb{R}^{m_{\text{mod}}} \;\middle|\; \langle \alpha, 1 \rangle = 1, \alpha \geq 0 \right\} .$$

It should be noted that (3) can be solved by the nonlinear programming problem

$$\min_{x,v} \left\{ v \middle| \; \hat{J}(x) \leq 1v;\; F(x) = 0 \right\} , \tag{4}$$

where $v \in \mathbb{R}^1$. The following two results are used to introduce the dual approach to this problem.

Lemma (1). Problem (3) is equivalent to

$$\min_{x} \max_{\alpha} \left\{ \langle \alpha, \hat{J}(x) \rangle \;\middle|\; F(x) = 0,\; \alpha \in \mathbb{R}_+^{m_{\text{mod}}} \right\} . \tag{5}$$

Proof. This result, initially proved by Medanic and Andjelic (1971; 1972) and also Cohen (1981), follows from the fact that the maximum of m_{mod} numbers is equal to the maximum of their convex combination. □

In Medanic and Andjelic (1971; 1972), the model is assumed to be linear, and the solution of (5) without the constraints $F(x) = 0$ is obtained using an iterative algorithm that projects α onto $\mathbb{R}_+^{m_{\text{mod}}}$. In Cohen (1981), the iterative nature of the projection is avoided by dispensing with the equality constraint in $\mathbb{R}_+^{m_{\text{mod}}}$ but including a normalization in a transformed objective function. Although the resulting objective function is not necessarily concave in the maximization variables, the algorithm proposed ensures convergence to the saddle point. The algorithm proposed in Cohen (1981) for nonlinear systems utilizes a simple projection procedure but is essentially first order.

Let α_* be the value of α that solves (5). It can be shown, by examining the first order conditions of (5) and (4), that α_* is also the shadow price associated with the inequality constraints in (4). An important feature of (5) that makes it preferable to (3) is that α^j can also be interpreted as the importance attached by the policy maker to the model $F^i(x) = 0$. *There may be cases in which the min-max solution α_* may be too extreme to implement. The policy maker may then wish to assign a value to α, in a neighbourhood of α_*, and determine a more acceptable policy by minimizing $\langle \alpha, \hat{J}(x) \rangle$, with respect to x, for the given α.* Another interpretation of (5) is in terms of the robust character of min-max policies. This is discussed in the following

Lemma:

Lemma (2). Let there exist a min-max solution to (5), denoted by (x_*, α_*), and let $\hat{J}$ and F be once differentiable at (x_*, α_*). Further, let strict complementarity hold for $\alpha \geq 0$ at this solution. Then, for $i, j, \ell \in \{1, 2, \ldots, m_{\text{mod}}\}$

(i)	$J^i(x_*) = J^j(x_*)$,	$\forall i, j \; (i \neq j)$	iff $\alpha_*^i, \alpha_*^j \in (0,1)$;
(ii)	$J^i(x_*) = J^j(x_*) > J^\ell(x_*)$,	$\forall i, j, \ell \; (\ell \neq i, j)$	iff $\alpha_*^\ell = 0$ and $\alpha_*^\ell = \alpha_*^j \in (0,1)$;
(iii)	$J^i(x_*) > J^j(x_*)$,	$\forall j, \; (j \neq i)$	iff $\alpha_*^i = 1$;
(iv)	$J^j(x_*) < J^j(x_*)$,	$\forall j, \; (j \neq i)$	iff $\alpha_*^i = 0$.

Proof. The necessary conditions for optimality of (5) are

$$\nabla_x \hat{J}(x_*)\, \alpha_* + \nabla_x F(x_*)\, \lambda_* = 0, \tag{6a}$$

$$\hat{J}(x_*) + \mu_* + 1\eta_* = 0, \tag{6b}$$

$$F(x_*) = 0, \tag{6c}$$

$$\langle 1, \alpha_* \rangle = 1\,, \quad \alpha_* \geq 0,$$

$$\langle \mu_*, \alpha_* \rangle = 0\,, \quad \mu_* \geq 0, \tag{7}$$

where λ_*, μ_*, η_* are the multipliers of $F(x) = 0, \alpha \geq 0$ and $\langle 1, \alpha \rangle = 1$, respectively.

Necessity in case (i) can be shown by considering (7), which, for α_*^i, $\alpha_*^j \in (0,1)$ yields $\alpha_*^i \mu_*^i = \alpha_*^j \mu_*^j = 0$, and then $\mu_*^i = \mu_*^j = 0$. Using (6) we have $J^i(x_*) = J^j(x_*)$. Sufficiency is established using $J^i(x_*) = J^j(x_*)$ and noting that

$$\eta_* = -\langle \hat{J}(x_*), \alpha_* \rangle\,.$$

Premultiplying the equality in (6b) by 1 and using this equality yields

$$0 = \langle 1, \hat{J}(x_*) \rangle + \langle 1, \mu_* \rangle + \langle 1, 1 \rangle\, \eta_* = \langle 1, \mu_* \rangle\,.$$

By (7), $\mu_* = 0$. Furthermore, strict complementarity implies that $\alpha_*^i \in (0,1), \; \forall i$.

Case (ii) can be shown by considering (7) for $\alpha_*^j, \alpha_*^j \in (0,1), \alpha_*^\ell = 0$. We have $\alpha_*^i \mu_*^i = \alpha_*^j\, \mu_*^j = \alpha_*^\ell \mu_*^\ell = 0$, thence $\mu_*^i = \mu_*^j = 0$ and, by strict complementarity, $\mu_*^\ell > 0$. From (6) we have

$$0 = J^m(x_*) + \eta_* + \mu_*^m\,, \quad m = i, j\,, \tag{8a}$$

$$0 = J^\ell(x_*) + \eta_* + \mu^\ell_* \,, \tag{8b}$$

and combining these yields

$$J^\ell(x_*) - J^m(x_*) = -\mu^\ell_* < 0; \; m = i, j \,.$$

For sufficiency, let $J^i(x_*) = J^j(x_*) > J^\ell(x_*)$. Combining (8) and using (7), we have

$$\alpha^\ell_*\Big(J^\ell(x_*) - J^m(x_*)\Big) = \alpha^\ell_*(\mu^m_* - \mu^\ell_*) = \alpha^\ell_* \mu^m_* \geq 0 \,.$$

Since $J^\ell(x_*) - J^m(x_*) < 0$, we have $\alpha^\ell_* = 0$. Given that $\alpha^\ell_* = 0$, $\forall \ell$, $J^\ell(x_*) < J^m(x_*)$, we can use (2) for those i, j for which $J^i(x_*) = J^j(x_*)$ to establish $\mu^i_* = \mu^j_* = 0$. By strict complementarity this implies that $\alpha^i_* \alpha^j_* \in (0, 1)$.

Case (iii) can be established noting that for $\alpha^j_* = 1$, we have $\mu^i_* = 0$, $\alpha^j_* = 0$, $\forall j \neq i$ and, by strict complementarity, $\mu^j_* > 0$. From (6) we thus obtain

$$J^j(x_*) - J^i(x_*) \leq \mu^i_* - \mu^j_* = -\mu^j_* < 0 \,.$$

Conversely, $J^i(x_*) > J^i(x_*)$ implies

$$\alpha^j_*\Big(J^j(x_*) - J^i(x_*)\Big) = \alpha^j_* \mu^i_* \geq 0$$

and thus $\alpha^j_* = 0$, $\forall j \neq i$. Case (iv) can be established as the converse of (iii). □

The above result illustrates the way in which α_* is related to $\hat{J}(x_*)$. When some of the elements of α_* are such that $\alpha^i_* \in (0, 1)$ for some $i \in M \subset \{1, 2, \ldots, m_{\text{mod}}\}$, it is shown that the $J^i(x_*)$ have the same value. *In this case, the optimal policy x_* yields the same objective function value whichever model happens to represent the economy.* Thus, x_* is a *robust policy*. In other circumstances, the policy maker is ensured that implementing x_* will yield an objective function value that is at least as good as the min-max optimum. This noninferiority of x_* may, on the other hand, amount to a cautious approach with high political costs. The policy maker can, in such circumstances, use α_* as a guide and seek in its neighbourhood a slightly less cautious scheme that is politically more acceptable. As mentioned above, this can be done by minimizing $\langle \alpha, \hat{J}(x) \rangle$ for a given value of α.

In a numerical example of the min-max approach (5), two models of the UK economy have been considered. One of these is the HM Treasury model (α^1) and the other is the NIESR model (α^2). The min-max solution is found to be $\alpha^1_* = .6$ and $\alpha^2_* = .4$. This is discussed further in Section 5.

In the algorithm discussed in the next section, a stepsize strategy is described that directly aims at measuring progress towards the min-max solution. The algorithm defines the direction of progress as a quasi-Newton step obtained from a quadratic subproblem. An augmented Lagrangian function is defined, and a procedure is formulated for determining the penalty parameter. The growth in the penalty parameter is required only to ensure a descent property. It is shown in Rustem (1992) that this penalty parameter does not grow indefinitely.

3. MIN-MAX ALGORITHM FOR RIVAL MODELS

Let the Lagrangian function associated with (2) be given by

$$L(x,\alpha,\lambda,\mu,\eta) = \langle \hat{J}(x),\alpha\rangle + \langle F(x),\lambda\rangle + \langle \alpha,\mu\rangle + (\langle 1,\alpha\rangle - 1)\,\eta\,, \tag{9}$$

where $\lambda \in \mathbb{R}^{m\times T}$, $\mu \in \mathbb{R}_+^{m_{\text{mod}}} = \{\mu \in \mathbb{R}^{m_{\text{mod}}} \mid \mu \geq 0\}$ and $\eta \in \mathbb{R}^1$ are the multipliers associated with $F(x) = 0$, $\alpha \geq 0$ and $\langle 1,\alpha\rangle = 1$, respectively. The characterization of the min-max solution of (2) as a saddle point requires the relaxation of convexity assumptions (see Demyanov and Malomezov, 1974; Cohen, 1981). In order to achieve this characterization, we modify (9) by augmenting it with a penalty function. Hence, we define the augmented Lagrangian by

$$L^a(x,\alpha,\lambda,\mu,\eta,c) = L(x,\alpha,\lambda,\mu,\eta) + \frac{c}{2} < F(x),F(x)\rangle\,, \tag{10}$$

where the scalar $c \geq 0$ is the penalty parameter.

In nonlinear programming algorithms, the penalty parameter c is either taken as a constant, is increased by a prefixed rate, or is adapted as the algorithm progresses. Specific examples of the adaptive strategy are Biggs (1974), Polak and Tits (1981), Polak and Mayne (1981). In this section, we also adopt such a strategy. However, we depart from the other works in adjusting c to ensure that the direction of search is a descent direction for the penalty function that regulates the stepsize strategy (14) below (Rustem, 1992; Lemmas 3.2 and 3.4). This approach is an extension of a strategy for nonlinear programming discussed in Rustem (1986, 1993).

Let $H(\cdot)$ H$(\cdot)$ denote the Hessians of L and L^a, with respect to x, evaluted at $(\cdot)$, respectively, and define the matrix

$$\nabla F(x) \equiv \{\nabla_x F^1(x),\ldots,\nabla_x F^e(x)]\,.$$

Sometimes, $\nabla F(x)$ evaluated at x_k will be denoted by ∇F_k, and $F(x_k)$ will be denoted by F_k. Thus, a local linearization of $F(x)$ at x_k can be written as

$$F(x) = F_k + \nabla F_k^T[x - x_k]\,.$$

Assumption (1). The columns of ∇F_k are assumed to be linearly independent. □

This assumption is used to simplify the quadratic subproblem used in the algorithm below for solving (2) and to ensure that the system $F_k + \nabla F_k^T[x - x_k]$ has a solution, $\forall x_k$. This assumption can be relaxed, but only by increasing the complexity of the quadratic subproblem.

Consider the objective function

$$\mathcal{F}(x,\alpha) = \langle \alpha, \hat{J}(x)\rangle$$

and its linear approximation, with respect to x, at a point x_k,

$$\mathcal{F}_k(x,\alpha) = \langle \alpha, \hat{J}(x_k) + \nabla\hat{J}(x_k)^T\,(x - x_k)\rangle\,, \tag{11a}$$

where

$$\nabla \hat{J}(x) = [\nabla J^1(x), \ldots, \nabla J^m(x)] .$$

We shall sometimes denote $\hat{J}(x)$ and $\nabla \hat{J}(x)$, evaluated at x_k, by $\hat{J}_k$ and $\nabla \hat{J}_k$, respectively. Thus, for $d = x - x_k$, (11,a) can be written as

$$\mathcal{F}_k(x_k + d, \alpha) = \langle \alpha, \hat{J}_k + \nabla \hat{J}_k^T d \rangle . \tag{11b}$$

The quadratic objective function used to compute the direction of progress is given by

$$q_k(x, \alpha, c) = \mathcal{F}_k(x, \alpha) + c\langle \nabla F_k F_k, (x - x_k) > + \tfrac{1}{2}\langle x - x_k, \hat{H}_k(x - x_k)\rangle$$

or, alternatively, by

$$q_k(d, \alpha, c) = \mathcal{F}_k(x_k + d, \alpha) + c\langle \nabla F_k F_k, d\rangle + \tfrac{1}{2}\langle d, \hat{H}_k d\rangle .$$

The matrix $\hat{H}_k$ is a symmetric positive semi-definite[1] approximation to the Hessian

$$H_k = \sum_{i=1}^{m} \alpha_k^i \nabla^2 J^i(x_k) + \sum_{j=1}^{e} \lambda_k^j \nabla^2 F^j(x_k) + c \nabla F_k \nabla F_k^T \tag{12}$$

The second derivatives due to the penalty term in the augmented Lagrangian (i.e. $c \sum_{j=1}^{e} \nabla^2 F^j(x_k) F^j(x_k)$) are not included in (12). The reason for this is discussed in Rustem (1992). Furthermore, since $F^j(x_*) = 0$ at the solution x_*, ignoring this term does not affect the asymptotic properties of the algorithm. The values α_k and λ_k are given by the solution to the quadratic subproblem in the previous iteration. The direction of progress at each iteration of the algorithm is determined by the quadratic subproblem

$$\min_d \max_\alpha \left\{ q_k(d, \alpha, c_k) \;\middle|\; \nabla F_k^T d + F_k = 0,\ \alpha \in \mathbb{R}_+^{m_{\text{mod}}} \right\} . \tag{13a}$$

Since the min-max subproblem is more complex, we also consider the quadratic programming subproblem

$$\min_{d,v} \left\{ v + c_k \langle \nabla F_k F_k, d\rangle + \tfrac{1}{2}\langle d, \hat{H}_k d\rangle \;\middle|\; \nabla F_k^T d + F_k = 0,\ \nabla \hat{J}_k^T d + \hat{J}_k \le 1v \right\} . \tag{13b}$$

The two subproblems are equivalent, but (13,b) involves fewer variables. It is shown below that the multipliers associated with the inequalities are the values α and that the solution of either subproblem satisfies common convergence properties.

Let the value of (d, α, v) solving (13) be denoted by $(d_k, \alpha_{k+1}, v_{k+1})$. The stepsize along d_k is defined using the equivalent min-max formulation (3). Thus, consider the function

[1] i.e. $\langle v, \hat{H}_k v\rangle \ge 0$, for all $v \ne 0$.

$$\psi(x) = \max_{i \in \{1,2,\ldots,m_{\text{mod}}\}} \left\{ J^i(x) \right\}$$

and

$$\psi_k(x) = \max_{i \in \{1,2,\ldots,m_{\text{mod}}\}} \left\{ J^i(x_k) + \langle \nabla J^i(x_k), \, x - x_k \rangle \right\} .$$

Let $\psi_k(x_k + d_k)$ be given by

$$\psi_k(x_k + d_k) = \max_{i \in \{1,2,\ldots,m_{\text{mod}}\}} \left\{ J^i(x_k) + \langle \nabla J^i(x_k), d_k \rangle \right\} .$$

The stepsize strategy determines τ_k as the largest value of $\tau = (\gamma)^j$, $\gamma \in (0,1)$, $j = 0, 1, 2, \ldots$ such that x_{k+1} given by

$$x_{k+1} = x_k + \tau_k d_k$$

satisfies the inequality

$$\psi(x_{k+1}) + \frac{c_{k+1}}{2} \langle F_{k+1}, F_{k+1} \rangle - \psi(x_k) - \frac{c_{k+1}}{2} \langle F_k, F_k \rangle \leq \rho \tau_k \Phi(d_k, c_{k+1}) , \tag{14a}$$

where $\rho \in (0,1)$ is a given scalar and

$$\Phi(d_k, c_{k+1}) = \psi_k(x_k + d_k) - \psi(x_k) + c_{k+1} \langle F_k, \nabla F_k^T d_k \rangle + \tfrac{1}{2} \langle d_k, \hat{H}_k d_k \rangle . \tag{14b}$$

The stepsize τ_k determined by (14) basically ensures that x_{k+1} simultaneously reduces the main objective and maintains or improves the feasibility with respect to the constraints. The penalty term used to measure this feasibility is quadratic and consistent with the augmented Lagrangian (10). It is shown in Rustem (1992; Theorem 4.1) that (14) can always be fulfilled by the algorithm.

The determination of the penalty parameter c is an important aspect of the algorithm. This is discussed in the following description:

The Algorithm

Step 0: Given $x_0, c_0 \in [0, \infty)$, and small positive numbers $\delta, \rho, \varepsilon, \gamma$ such that $\delta \in (0, \infty)$, $\rho \in (0,1)$, $\varepsilon \in (0, \frac{1}{2}]$, $\gamma \in (0,1)$, $\hat{H}_0$, set $k = 0$.

Step 1: Compute $\nabla \hat{J}_k$ and ∇F_k. Solve the quadratic subproblem (13) (choosing (13,a) or (13,b) defines a particular algorithm) to obtain d_k, α_{k+1}, and the associated multiplier vector λ_{k+1}. In (13,a), we also compute μ_{k+1}, η_{k+1} and in (13,b) we also compute v_{k+1}.

Step 2: Test for optimality: If optimality is achieved, *stop*. Else go to *Step 3*.

Step 3: If

$$\psi_k(x_k + d_k) - \psi(x_k) - c_k \langle F_k, F_k \rangle + (\varepsilon + \tfrac{1}{2}) \langle d_k, \hat{H}_k d_k \rangle \leq 0 ,$$

then $c_{k+1} = c_k$. Else set

$$c_{k+1} = \max \left\{ \frac{\psi_k(x_k + d_k) - \psi(x_k) + (\varepsilon + \frac{1}{2})\langle d_k, \hat{H}_k d_k \rangle}{\langle F_k, F_k \rangle}, \; c_k + \delta \right\} . \tag{15}$$

Step 4: Find the smallest nonnegative integer j_k such that $\tau_k = \gamma^{j_k}$ with $x_{k+1} = x_k + \tau_k d_k$ such that the inequality (14) is satisfied.

Step 5: Update $\hat{H}_k$ to compute $\hat{H}_{k+1}$, set $k = k + 1$ and go to Step 1.

In Step 3, the penalty parameter c_{k+1} is adjusted to ensure that progress towards feasibility is maintained. In particular, c_{k+1} is chosen to make sure that the direction d_k computed by the quadratic subproblem is a descent direction for the penalty function $\psi(x_k) - \frac{c_{k+1}}{2}\langle F_k, F_k \rangle$.

In Rustem (1992), it is shown that d_k is a descent direction, that c_k determined by (15) is not increased indefinitely, that the algorithm converges to a local solution of the min-max problem, that the stepsize stepsize τ_k converges to unity, and that the local convergence rate near the solution is Q- or two-step Q-superlinear, depending on the accuracy of the approximate Hessian, $\hat{H}_k$.

4. NUMERICAL EXPERIMENTS

In this section, we illustrate the behaviour of the method with a few test examples. The objective is to highlight the characteristics of the algorithm along with certain properties of min-max problems. Specifically, we show the attainment of unit stepsizes ($\tau_k = 1$), the way in which the penalty parameter c_k achieves the constant value c_*, and the numbers of iterations and function evaluations needed to reach the solution in each case. The attainment of a constant penalty parameter is important for numerical stability. The achievement of unit steps is important in ensuring rapid superlinear convergence (Rustem, 1992).

We also show the progress of the algorithm towards the min-max solution, which exhibits certain robustness characteristics predicted by theory. As discussed in Lemma 2, if the min max over three functions J^1, J^2 and J^3 is being computed, then, at the solution, $J^1 = J^2 > J^3$ iff $\alpha_1, \alpha_2 \in (0, 1]$ and $\alpha_3 = 0$ or $J^1 > J^2 \geq J^3$ iff $\alpha_1 = 1$ and $\alpha_2 = \alpha_3 = 0$.[2] Lemma 2 states this in greater generality, and the examples illustrate it. Since α is chosen to maximize the Lagrangian (9), the solution

[2] Suppose that the state of the world is described by, say, three rival theories one of which is to turn out to be the actual state. With $J^1 = J^2 > J^3$, at the min-max solution, the decision maker need not care, as far as the objective function values are concerned, if the actual state turns out to be J^1 or J^2. If it is J^3, then the decision maker is better off. The Lagrange multiplier vector α indicates this in the min-max formulation (4) and the associated subproblem (13,b). The robustness aspect is underlined by Lemma 2.

can be seen as a robust optimum in the sense of a worst-case design problem. The figures describing the convergence of the algorithms also illustrate the process of convergence of the objective functions to the min-max optima.

We consider six test examples. Three of these are unconstrained min-max problems in which we study the achievement of unit steplengths, and three are constrained problems in which we study both the achievement of *unit stepsizes* and a *constant penalty parameter* value c_*. The approximate Hessian computation uses the BFGS updating formula and, for constrained problems, its modification discussed in Powell (1978). The Hessian approximation is done on the second derivative terms arising from the Lagrangian (i.e. the first two terms on the right of (12)) whereas the exact value for the term $c_k N_k N_k^T$ is used. The other parameters of the algorithms are set at $\delta = 0.01$, $\rho = 0.1$, $\gamma = 0.1$.

Example 1. (Charalambous and Bandler, 1976)

$$\left\{ J^1(x) = x_1^2 + x_2^4;\ J^2(x) = (2 - x_1)^2 + (2 - x_2)^2;\ J^3(x) = 2\ \exp(x_2 - x_1) \right\}$$

Initial values: $x_0^T = [1, -0.1]$; $\alpha_0^T = [\frac{1}{3}, \frac{1}{3}, \frac{1}{3}]$

	Hessian evaluation scheme	
	Direct	BFGS
x_*	1.1390376520	1.1390376520
	0.8995599640	0.8995599340
α_*	0.4304811740	0.4304811740
	0.5695188260	0.5695188360
	0.0	0.0
$J^1(x_*)$	1.9522244939	1.9522244939
$J^2(x_*)$	1.9522244939	1.9522244939
$J^3(x_*)$	1.5740776268	1.5740776268
$k_\tau \big\vert \tau_k = 1;\ \forall k \geq k_\tau$ [3]	1	1
No. of J evaluations	5	7
No. of iterations	5	7

The same example was also computed with the initial value of x changed to $[0, 0]$. All the results are identical except that the algorithms took 6 iterations and function evaluations for the direct Hessian case and 8 iterations and function evaluations for the Hessian approximated by the BFGS formula.

[3] k_τ denotes the iteration at which $\tau_k = 1$ was reached and unit stepsizes were maintained $\forall k \geq k_\tau$.

Example 2. (Charalambous and Bandler, 1976)

$$\left\{ J^1(x) = x_1^4 + x_2^2;\ J^2(x) = (2 - x_1)^2 + (2 - x_2)^2;\ J^3(x) = 2\ \exp(x_2 - x_1) \right\}$$

Initial values: $x_0^T = [1, -0.1]$; $\alpha_0^T = [\frac{1}{3}, \frac{1}{3}, \frac{1}{3}]$

	Hessian evaluation scheme	
	Direct	BFGS
x_*	1.0	1.0
	1.0	1.0
α_*	0.33333333	0.33333333
	0.5	0.5
	0.16666667	0.16666667
$J^1(x_*)$	2.0	2.0
$J^2(x_*)$	2.0	2.0
$J^3(x_*)$	2.0	2.0
$k_\tau \,\vert\, \tau_k = 1;\ \forall k \geq k_\tau$	1	1
No. of J evaluations	5	5
No. of iterations	5	5

The same example was also computed from the initial value $[0, 0]$. The same result in reached in 7 iterations for both Hessian evaluation schemes and unit stepsizes are again achieved at iteration 1.

Example 3. (Polak, Mayne and Higgins, 1986)

$$\left\{J^1(x) = \exp\left(\frac{x_1^2}{1000} + (x_2 - 1)^2\right);\ J^2(x) = \exp\left(\frac{x_1^2}{1000} + (x_2 + 1)^2\right)\right\}$$

Initial values: $x_0^T = [50.0, 0.05]$; $\alpha_0^T = [0.5, 0.5]$

	Hessian evaluation scheme	
	Direct	BFGS
x_*	8.1917×10^{-9}	2.5711×10^{-9}
	1.09957×10^{-15}	6.7297×10^{-21}
α_*	0.5	0.5
	0.5	0.5
$J^1(x_*)$	2.718281828	2.718281828
$J^2(x_*)$	2.718281828	2.718281828
$k_\tau \mid \tau_k = 1;\ \forall k \geq k_\tau$	1	1
No. of J evaluations	8	11
No. of iterations	8	11

If the algorithm is started from the initial point [1.0, 1.1], the same result is obtained in 7 (direct) and 11 (BFGS) iterations.

Example 4. (Conn, 1979)

$$\left\{J^1(x) = x_1^2 + x_2^2;\ J^2(x) = (2 - x_1)^2 + (2 - x_2)^2;\ J^3(x) = 2\ \exp(x_2 - x_1)\right\}$$

subject to the constraints

$$\left\{F^1(x) = x_1 + x_2 - 2;\ F^2(x) = -x_1^2 - x_2^2 + 2.25\right\}.$$

This example is solved for different initial values and for two different initial penalty parameters $c_0 = 1.0$ and $c_0 = 10.0$. Other initial values for all cases are $\alpha_0^T = [\frac{1}{3}, \frac{1}{3}, \frac{1}{3}]$ and the Lagrange multiplier estimates $\lambda_0^T = [\frac{1}{3}, \frac{1}{3}]$.
Initial value $x_0^T = [2.1, 1.9]$

	$c_0 = 1.0$ Hessian evaluation scheme		$c_0 = 10.0$ Hessian evaluation scheme	
	Direct	BFGS	Direct	BFGS
x_*	1.353553391	1.353553391	1.353553390	1.353553391
	0.646446609	0.646446609	0.646446609	0.646446609
α_*	0.0	0.0	0.0	0.0
	1.0	1.0	1.0	1.0
	0.0	0.0	0.0	0.0
λ_*	4.000000015	4.0	4.000000008	3.999999997
	1.000000008	0.999999987	1.000000004	0.999999976
$J^1(x_*)$	2.250000004	2.250000002	2.250000002	2.250000002
$J^2(x_*)$	2.250000004	2.250000002	2.250000002	2.250000002
$J^3(x_*)$	0.986137337	0.986137381	0.9861378	0.9861378
c_* (final)	1780.693576	2.488613732	31.86545192	46.69260520
$k_\tau \big\vert \tau_k = 1;\ \forall k \geq k_\tau$	4	3	3	3
$k_* \big\vert c_k = c_*;\ \forall k \geq k_*$ [4]	7	4	3	5
No. of fn. eval.	15	12	12	12
No. of iterations	10	9	9	9

[4] Iteration number at which c_k reached a value c_* such that $c_k = c_*$ $\forall k$ after this iteration.

Initial value $x_0^T = [1.9, 2.1]$ from which the algorithm converges to a local solution.

	$c_0 = 1.0$ Hessian evaluation scheme		$c_0 = 10.0$ Hessian evaluation scheme	
	Direct	BFGS	Direct	BFGS
x_*	0.646446609	0.646446609	0.646446609	0.646446609
	1.353553390	1.353553390	1.353553390	1.353553390
α_*	0.0	0.0	0.0	0.0
	0.0	0.0	0.0	0.0
	1.0	1.0	1.0	1.0
λ_*	11.47275085	11.47275085	11.47275085	11.47275085
	5.73637543	5.73637542	5.73637543	5.73637542
$J^1(x_*)$	2.250000002	2.250000002	2.250000002	2.250000002
$J^2(x_*)$	2.250000002	2.250000002	2.250000002	2.250000002
$J^3(x_*)$	4.056229975	4.056229975	4.056229975	4.056229975
c_* (final)	6601.845484	2.442484228	79.34209	44.1686
$k_\tau \middle\| \tau_k = 1;\ \forall k \geq k_\tau$	4	3	3	3
$k_* \middle\| c_k = c_*;\ \forall k \geq k_*$ [4]	7	2	5	5
No. of fn. eval.	28	12	12	12
No. of iterations	10	9	9	9

Initial value $x_0^T = [4, 2]$.

	$c_0 = 1.0$ Hessian evaluation scheme		$c_0 = 10.0$ Hessian evaluation scheme	
	Direct	BFGS	Direct	BFGS
x_*	1.353553392	1.353553391	1.353553392	1.353553391
	0.646446609	0.646446609	0.646446609	0.646446609
α_*	0.0	0.0	0.0	0.0
	1.0	1.0	1.0	1.0
	0.0	0.0	0.0	0.0
λ_*	4.000000764	3.999999990	4.0	4.0
	1.000000386	0.999999994	1.0	1.0
$J^1(x_*)$	2.250000002	2.250000002	2.25	2.250000002
$J^2(x_*)$	2.250000002	2.250000002	2.25	2.250000002
$J^3(x_*)$	0.986137356	0.986137356	0.9861356	0.9861356
c_* (final)	1.0	1.0	1.0	12.5924227
$k_\tau \Big\vert \tau_k = 1;\ \forall k \geq k_\tau$	1	1	1	1
$k_* \Big\vert c_k = c_*;\ \forall k \geq k_*$ [4]	1	1	1	1
No. of fn. eval.	8	8	8	8
No. of iterations	8	8	8	8

Initial value $x_0^T = [2, 4]$ from which the algorithm converges to a local solution.

	$c_0 = 1.0$ Hessian evaluation scheme		$c_0 = 10.0$ Hessian evaluation scheme	
	Direct	BFGS	Direct	BFGS
x_*	0.646446609	0.646446609	0.646446609	0.646446609
	1.353553390	1.353553390	1.353553390	1.353553390
α_*	0.0	0.0	0.0	0.0
	0.0	0.0	0.0	0.0
	1.0	1.0	1.0	1.0
λ_*	11.47275085	11.47275085	11.47275085	11.47275085
	5.73637543	5.73637542	5.73637543	5.73637542
$J^1(x_*)$	2.25	2.25	2.250000002	2.250000002
$J^2(x_*)$	2.25	2.25	2.250000002	2.250000002
$J^3(x_*)$	4.056229963	4.056229964	4.056229963	4.056223007
c_* (final)	1.21109174	2.094947393	24.8366	10.165631
$k_\tau \big\vert \tau_k = 1;\ \forall k \geq k_\tau$	3	2	1	1
$k_* \big\vert c_k = c_*;\ \forall k \geq k_*$ [4]	4	4	4	3
No. of fn. eval.	12	10	8	8
No. of iterations	10	9	8	8

Example 5. The min-max formulation of the Rosen-Suzuki problem (Conn, 1979)

$$J^1(x) = x_1^2 + x_2^2 + 2x_3^2 + x_4^2 - 5x_1 - 5x_2 - 21x_3 + 7x_4\,,$$

$$J^2(x) = -9x_1^2 - 9x_2^2 - 8x_3^2 - 9x_4^2 - 15x_1 + 5x_2 - 31x_3 + 17x_4 + 80\,,$$

$$J^3(x) = 11x_1^2 + 11x_2^2 + 12x_3^2 + 11x_4^2 + 5x_1 - 15x_2 - 11x_3 - 3x_4 - 80\,,$$

subject to the constraints

$$F^1(x) = -x_1^2 - 2x_2^2 - x_3^2 - 2x_4^2 + x_1 + x_4 + 10\,,$$

$$F^2(x) = -2x_1^2 - x_2^2 - x_3^2 - 2x_1 + x_2 + x_4 + 5\,.$$

This example is solved for different initial penalty parameters $c_0 = 1.0$ and $c_0 = 10.0$. Initial values $x_0^T = [0, 1, 1, 0]$; $\alpha_0^T = [\frac{1}{3}, \frac{1}{3}, \frac{1}{3}]$; $\lambda_0^T = [\frac{1}{3}, \frac{1}{3}]$.

	$c_0 = 1.0$ Hessian evaluation scheme		$c_0 = 10.0$ Hessian evaluation scheme	
	Direct	BFGS	Direct	BFGS
x_*	-2.3606×10^{-9}	1.48215×10^{-14}	-2.3606×10^{-7}	1.02673×10^{-12}
	1.00000003	1.0	1.00000003	1.0
	2.00000016	2.0	2.00000016	2.0
α_*	0.0	0.0	0.0	0.0
	0.45	0.45	0.45	0.45
	0.55	0.55	0.55	0.55
λ_*	-8.0938×10^{-9}	1.04033×10^{-9}	-8.0911×10^{-9}	1.45367×10^{-9}
	2.0000006	2.0	2.0000006	2.0
$J^1(x_*)$	-44.0	-44.0	-44.0	-44.0
$J^2(x_*)$	-44.0	-44.0	-44.0	-44.0
$J^3(x_*)$	-44.0	-44.0	-44.0	-44.0
c_* (final)	1.0	1.0	10.0	10.0
$k_\tau \mid \tau_k = 1;\ \forall k \geq k_\tau$	1	1	1	15
$k_* \mid c_k = c_*;\ \forall k \geq k_*$ [4]	1	1	1	1
No. of fn. eval.	20	9	20	26
No. of iterations	20	9	20	19

Example 6. Two constraints imposed on Example 2.

$$\left\{J^1(x) = x_1^4 + x_2^2;\ J^2(x) = (2 - x_1)^2 + (2 - x_2)^2;\ J^3(x) = 2\ \exp(x_2 - x_1)\right\}$$

subject to the constraints

$$\left\{F^1(x) = x_1^2 - x_2^2;\ F^2(x) = -2x_1^3 - x_2^2\right\}.$$

This example is solved for different initial penalty parameters $c_0 = 1.0$ and $c_0 = 10.0$. Initial values $x_0^T = [0, 1]$; $\alpha_0^T = [\frac{1}{3}, \frac{1}{3}, \frac{1}{3}]$.

	$c_0 = 1.0$ Hessian evaluation scheme		$c_0 = 10.0$ Hessian evaluation scheme	
	Direct	BFGS	Direct	BFGS
x_*	1.000000069	1.0000003456	1.0000003456	1.0000003456
	1.000000069	1.0000003456	1.0000003456	1.0000003456
α_*	1.0	1.0	1.0	1.0
	0.0	0.0	0.0	0.0
	0.0	0.0	0.0	0.0
λ_*	2.500000147	2.5	2.500000147	2.5
	-1.500000112	-1.499995219	-1.500000112	-1.49999811429
$J^1(x_*)$	2.0000008294	2.00000004147	2.00000004147	2.00000004147
$J^2(x_*)$	1.9999994471	1.9999997235	1.9999997235	1.9999997235
$J^3(x_*)$	2.0	2.0	2.0	2.0
c_* (final)	6.381975949	137.83894126	10.431886987	545.15388777
$k_\tau \mid \tau_k = 1;\ \forall k \geq k_\tau$	5	5	5	5
$k_* \mid c_k = c_*;\ \forall k \geq k_*$ [4]	5	3	5	3
No. of fn. eval.	23	23	23	23
No. of iterations	20	20	20	20

Figures 1–3 below describe the behaviour of the objective functions as the algorithm proceeds to the optimum. The figures for the constrained examples 4–6 correspond to the results for the initial penalty parameter $c_0 = 10$.

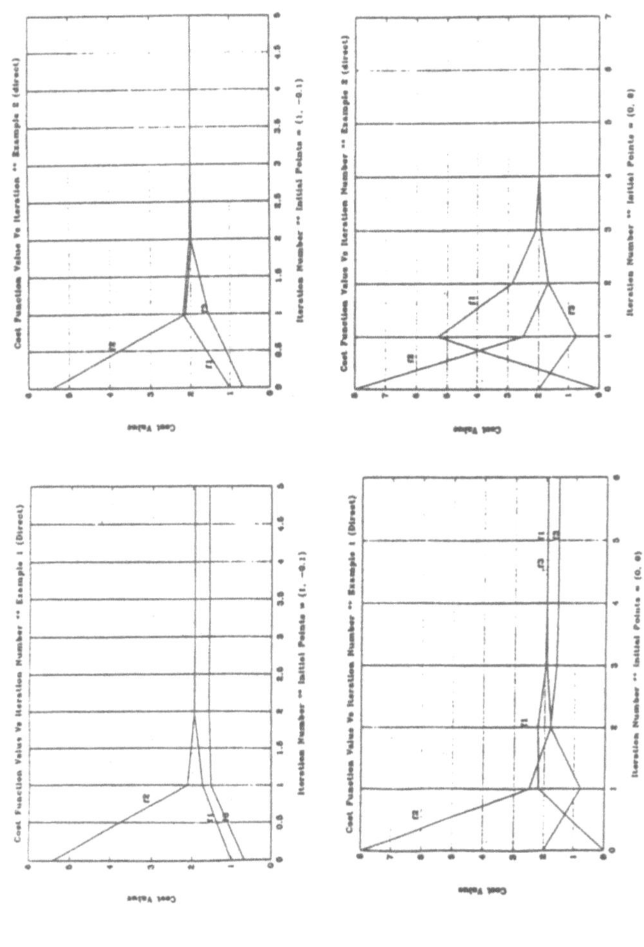

Fig. 1. The convergence of the objective functions vs iteration number for examples 1–2 in Section 4. "Direct" signifies the direct (exact, as opposed to approximate BFGS) evaluation of Hessians.

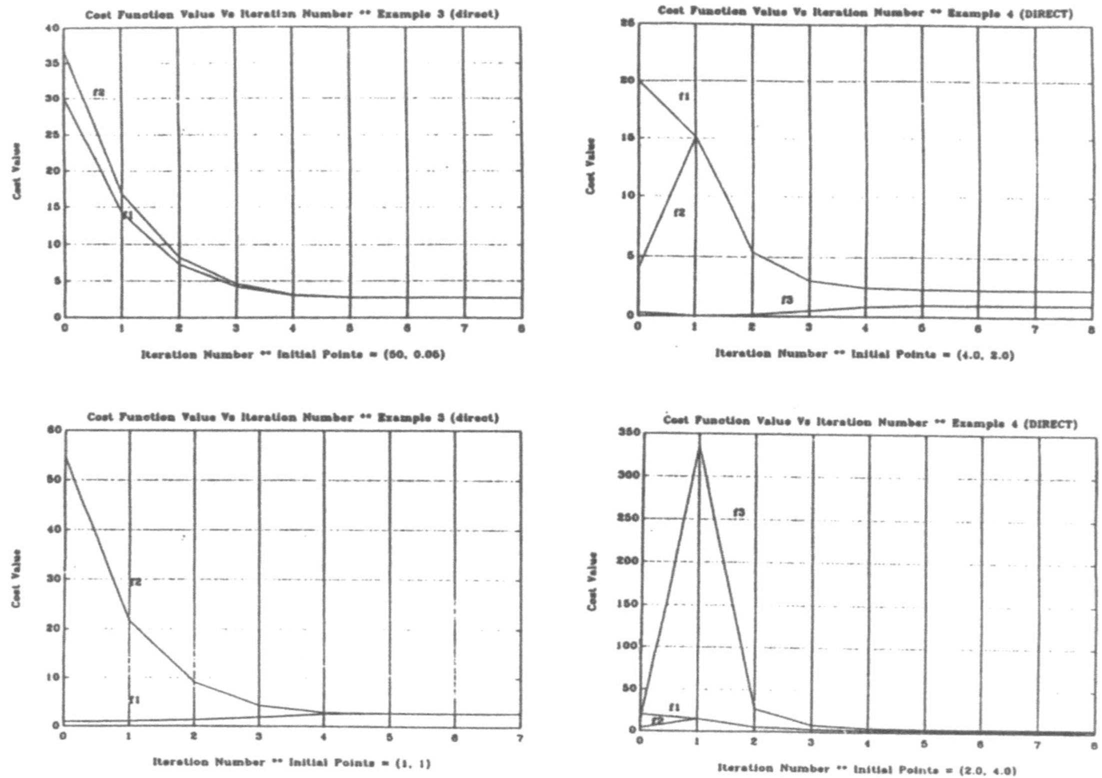

Fig. 2. The convergence of the objective functions vs iteration number for examples 3–4 in Section 4. "Direct" signifies the direct (exact, as opposed to approximate BFGS) evaluation of Hessians.

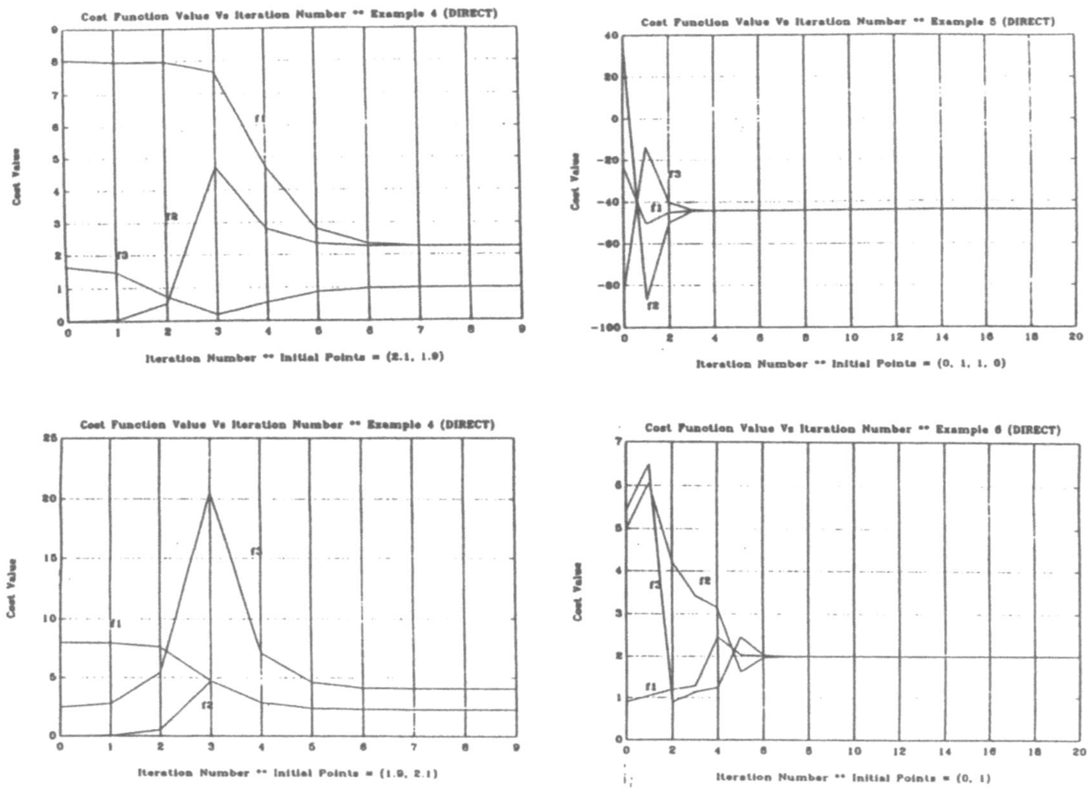

Fig. 3. The convergence of the objective functions vs iteration number for examples 4–6 in Section 4. "Direct" signifies the direct (exact, as opposed to approximate BFGS) evaluation of Hessians.

5. THE UNCONSTRAINED CASE FOR RIVAL ECONOMIC MODELS

The constrained min-max problem (2) may also be formulated as an unconstrained problem if it is possible to use the model i to eliminate Y^i and express (2) in terms of U only. However, this may not be always feasible, if for example there are constraints other than those of the model or if it is not possible simply to eliminate the model due to complex perfect-foresight conditions. Given the solution software that accompanies most models, it is possible to evaluate Y^i for every value of U. We thus have a numerical representation of the model in the form $Y^i = g^i(U)$. The model derivative information dY^i/dU can also be obtained numerically using this approach. The advantages of this approach are that the unconstrained problem is solved, thereby avoiding potential complications due to constraints, and the dimensionality of the problem is reduced to that of U. A disadvantage of the approach is that the model solution is repeatedly computed to evaluate Y^i for every U considered by the algorithm. By contrast, the constrained algorithm *simultaneously* converges to an optimal and feasible $[Y^i, U]$ pair. Another disadvantage is that the derivative information dY^i/dU inevitably involves numerical inaccuracies and that may affect convergence.

Once Y^i has been eliminated using the model, the min-max problem (5) can be written as

$$\min_{U} \max_{\alpha^i} \left\{ \sum_1^m \alpha^i f^i(U) \,\middle|\, 0 \leq \alpha^i \leq 1; \sum_1^m \alpha^i = 1;\ i = 1, \ldots, m \right\}, \tag{16}$$

where $f^i(U)$ is the reduced objective function corresponding to the ith model after Y^i have been evaluated and eliminated from J^i.

The algorithm for solving (16) is essentially a simpler version of that for the constrained case when the penalty parameter $c_k = 0$, $\forall k$. A sequence $\{U_k\}$ is generated so that, starting from a given initial policy vector, each member of the sequence is defined by

$$U_{k+1} = U_k + \tau_k d_k\ , \tag{17}$$

where d_k is the direction of search computed by solving the quadratic subproblem (18) below, and τ_k is the stepsize.

Consider the quadratic programming problem

$$\min_{d,v} \left\{ v + \tfrac{1}{2} d^T \hat{H}_k d \,\middle|\, \langle \nabla f_i(U_k), d \rangle + f_i(U_k) \leq v;\ i = 1, \ldots, m \right\}. \tag{18}$$

Here, as in (13,b), v is a scalar, and d_k is the direction of search. $\hat{H}_k$ is a symmetric positive definite quasi-Newton approximation to the Hessian

$$H_k = \sum_1^m \alpha_k^i \nabla^2 f^i(U_k)\ ,$$

and $\nabla f^i(U_k)$ is the column vector denoting the gradient of f^i. The shadow prices or the Kuhn-Tucker multipliers of (18) give the values α_{k+1}^i, $i = 1, \ldots, m$.

In order to introduce the stepsize strategy τ, we consider the function

$$\psi(U) = \max_{i \in \{1,2,\ldots,m\}} \{f^i(U)\}$$

and

$$\psi_k(U) = \max_{i \in \{1,2,\ldots,m\}} \{f^i(U_k) + \langle \nabla f^i(U_k), U - U_k \rangle\} .$$

Let $\psi_k(U_k + d_k)$ be given by

$$\psi_k(U_k + d_k) = \max_{i \in \{1,2,\ldots,m\}} \{f^i(U_k) + \langle \nabla f^i(U_k), d_k \rangle\} .$$

The stepsize τ_k is the largest value of $\tau = (\gamma)^j, j = 0, 1, 2, \ldots$ such that U_{k+1} given by (17) satisfies the inequality

$$\psi(U_{k+1}) - \psi(U_k) \leq \rho \tau_k \Phi(d_k) ,$$

where $\rho \in (0, 1)$ is a given scalar and

$$\Phi(d_k) = \psi_k(U_k + d_k) - \psi(U_k) + \tfrac{1}{2} \langle d_k, \hat{H}_k d_k \rangle .$$

The unconstrained case has been considered for economic models in Becker *et al.* (1986), with the UK economy models of HM Treasury and NIESR as rivals, and in Karakitsos and Rustem (1991) where three small rival economic models are considered. In the two model case discussed by Becker *et al.* (1986), the robust characterization of min-max, summarized in Lemma 2, is illustrated by Figure 4. The vertical axis gives the objective function values versus changes in α on the horizontal axis. The min-max point is where the three curves meet. The concave curve is the plot of

$$\min_U \{\alpha f^{\text{HMT}}(U) + (1 - \alpha) f^{\text{NIESR}}(U)\} \tag{19}$$

for *given values of* $\alpha \in [0, 1]$. Its maximum corresponds to the value of α maximising the above function and thence to the min-max solution. The functions $f^{\text{HMT}}(U)$ and $f^{\text{NIESR}}(U)$ correspond to the reduced objective functions obtained after eliminating Y^{HMT} and Y^{NIESR} from (2), using their respective models. The two convex curves are the corresponding values of $f^{\text{HMT}}(U)$ and $f^{\text{NIESR}}(U)$ in (19). As α increases, $f^{\text{HMT}}(U)$ decreases, and as $(1 - \alpha)$ increases, $f^{\text{NIESR}}(U)$ decreases. The curves meet at $\alpha = .6$, which corresponds to the min-max value. At this point $f^{\text{HMT}}(U) = f^{\text{NIESR}}(U)$, as discussed in Lemma 2, and the policy corresponding to this point is model invariant.

The performance of the preceding unconstrained algorithm was studied in connection with the computations involving the three rival models in Karakitsos and Rustem (1991). Each model is a discrete-time, nonlinear dynamic system of 11 equations with perfect foresight and 2 policy instruments; the problem is solved over 5 periods. Thus, dim $(U) = 10$. This problem was solved numerous times, with different parameter specifications and initial points. Using a convergence criterion of 10^{-5} for the satisfaction of the optimality conditions, the algorithm converged within 11–38 iterations, and unit stepsize ($\tau_k = 1$) was achieved at iteration 7–8 in two

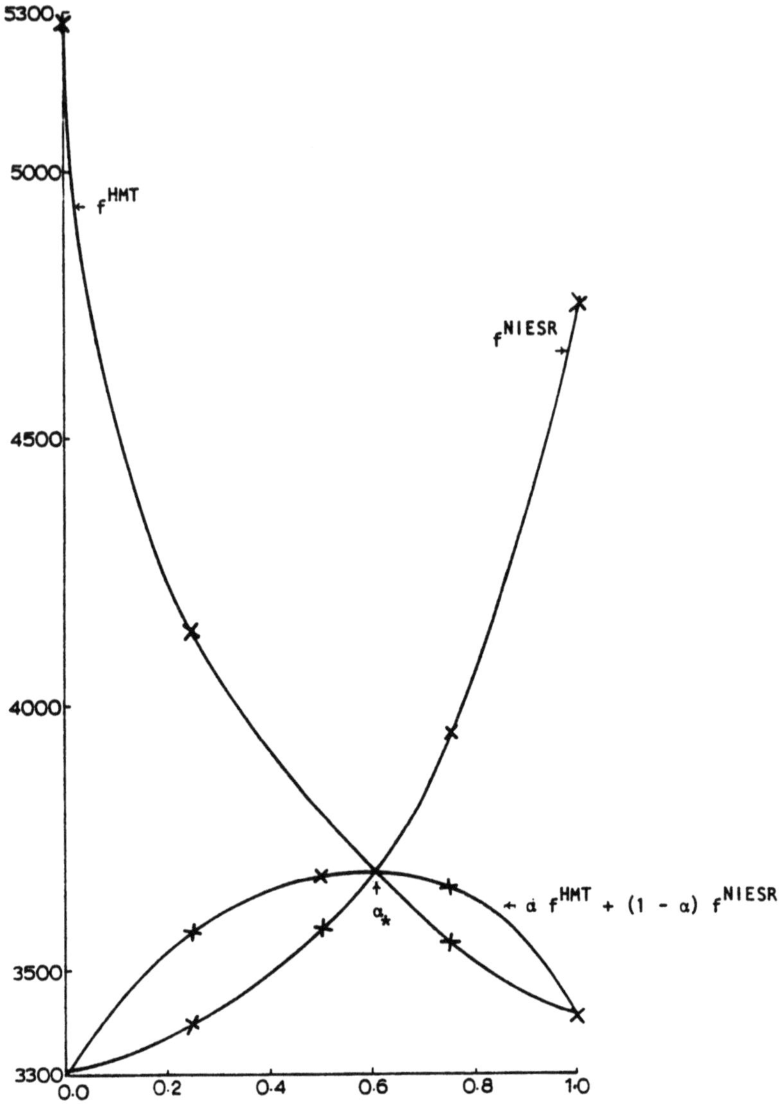

Fig. 4. The variation of $\min_U \{\alpha f^{HMT}(U) + (1-\alpha) f^{NIESR}(U)\}$ as α varies, $0 \leq \alpha \leq 1$, and the corresponding values of the individual components $f^{HMT}(U)$, $f^{NIESR}(U)$.

out of seven cases. As opposed to the examples in Section 4, any failure to achieve $\tau_k = 1$ could mostly be attributed to the errors in the numerical evaluation of the model and dY^i/dU. Such errors affect the accuracy of the Hessian approximation of f^i, which requires dY^i/dU, and $\tau_k = 1$ can be shown to depend on the accuracy of this approximation (see Rustem, 1992).

ACKNOWLEDGEMENTS

The valuable comments and suggestions of David Belsley are gratefully acknowledged.

REFERENCES

Becker, R.G., B. Dwolatzky, E. Karakitsos and B. Rustem (1986). "The Simultaneous Use of Rival Models in Policy Optimization", *The Economic Journal* **96**, 425–448.

Biggs, M.C.B. (1974). "The Development of a Class of Constrained Minimization Algorithms and their Application to the Problem of Power Scheduling", Ph.D. Thesis, University of London.

Charalambous, C. and J.W. Bandler (1976). "Nonlinear Minimax Optimization as a Sequence of Least *p*th Optimization with Finite Values of *p*", *International Journal of System Science* **7**, 377–391.

Charalambous, C. and A.R. Conn (1978). "An Efficient Algorithm to Solve the Min-Max Problem Directly", *SIAM J. Num. Anal.* **15**, 162–187.

Chow, G.C. (1979). "Effective Use of Econometric Models in Macroeconomic Policy Formulation" *in* "Optimal Control for Econometric Models" (S. Holly, B. Rustem, M. Zarrop, eds.), Macmillan, London.

Cohen, G. (1981). "An Algorithm for Convex Constrained Minmax Optimization Based on Duality", *Appl. Math. Optim.* **7**, 347–372.

Coleman, T.F. (1978). "A Note on 'New Algorithms for Constrained Minimax Optimization' ", *Math. Prog.* **15**, 239–242.

Conn, A.R. (1979). "An Efficient Second Order Method to Solve the Constrained Minmax Problem", Department of Combinatorics and Optimization, University of Waterloo, Report January.

Demyanov, V.F. and V.N. Malomezov (1974). "Introduction to Minmax", J. Wiley, New York.

Demyanov, V.F. and A.B. Pevnyi (1972). "Some Estimates in Minmax Problems", *Kibernetika* **1**, 107–112.

Dutta, S.R.K. and M. Vidyasagar (1977). "New Algorithms for Constrained Minmax Optimization", *Math. Prog.* **13**, 140–155.

Granger, C. and P. Newbold (1977). "Forecasting Economic Time Series", Academic Press, New York.

Han, S–P. (1978). "Superlinear Convergence of a Minimax Method", Dept. of Computer Science, Cornell University, Technical Report 78–336.

Han, S–P. (1981). "Variable Metric Methods for Minimizing a Class of Nondifferentiable Functions", *Mathematical Programming* **20**, 1–13.

Karakitsos, E. and B. Rustem (1991). "Min-Max Policy Design with Rival Models", *PROPE Discussion Paper* 116, Presented at the SEDC Meeting, Minnesota.

Lawrence, M.J., R.H. Edmunson and M.J. O'Connor (1986). "The Accuracy of Combining Judgemental and Statistical Forecasts", *Management Science* **32**, 1521–1532.

Makridakis, S. and R. Winkler (1983). "Averages of Forecasts: Some Empirical Results", *Management Science* **29**, 987–996.

Medanic, J. and M. Andjelic (1971). "On a Class of Differential Games without Saddle-point Solutions", *JOTA* **8**, 413–430.

Medanic, J. and Andjelic (1972). "Minmax Solution of the Multiple Target Problem", *IEEE Trans. AC–17*, 597–604.

Murray, W. and M.L. Overton (1980). "A Projected Lagrangian Algorithm for Nonlinear Minmax Optimization", *SIAM J. Sci. Stat. Comput.* **1**, 345–370.

Polak, E. and D.Q. Mayne (1981). "A Robust Secant Method for Optimization Problems with Inequality Constraints", *JOTA* **33**, 463–477.

Polak, E., D.Q. Mayne and J.E. Higgins (1988). "A Superlinearly Convergent Min-Max Algorithm for Min-Max Problems", *Memorandum No: UCB/ERL M86/103*, Berkely California.

Polak, E. and A.L. Tits (1980). "A Globally Convergent, Implementable Multiplier Method with Automatic Penalty Limitation", *Appl. Math. and Optimization* **6**, 335–360.

Powell, M.J.D. (1978). "A Fast Algorithm for Nonlinearly Constrained Optimization Problems", in G.A. Watson (Ed.), *Numerical Analysis, Lecture Notes in Mathematics*, 630, Springer-Verlag, Berlin-Heidelberg.

Rustem, B. (1986). "Convergent Stepsizes for Constrained Optimization Algorithms", *JOTA* **49**, 135–160.

Rustem, B. (1987). "Methods for the Simultaneous Use of Multiple Models in Optimal Policy Design", *in* "Developments in Control Theory for Economic Analysis", (C. Carraro and D. Sartore, eds.), Martinus Nijhoff Kluwer Publishers, Dordrecht.

Rustem, B. (1989). "A Superlinearly Convergent Constrained Min-Max Algorithm for Rival Models of the Same System", *Comp. Math. Applic.* **17**, 1305–1316.

Rustem, B. (1992). "A Constrained Min-Max Algorithm for Rival Models of the Same Economic System", *Mathematical Programming* **53**, 279–295.

Rustem, B. (1993). "Equality and Inequality Constrained Optimization Algorithms with Convergent Stepsizes", forthcoming *JOTA*.

PART THREE

Computational Techniques for Econometrics

WILLIAM L. GOFFE

Wavelets in Macroeconomics: An Introduction

ABSTRACT. Wavelets are a new method of spectral analysis that have attracted considerable attention in numerous fields. Unlike Fourier methods, wavelets are designed to analyze data that is nonstationary and subject to abrupt changes. Since macroeconomic data frequently contains these characteristics, wavelets appear to be a natural tool for studying macroeconomic time series. This paper first describes wavelets in an intuitive manner, and then explores their use on macroeconomic data. Initial results are encouraging and more research is in order.

1. INTRODUCTION

Wavelets, a method of signal analysis, have attracted considerable recent attention. The *IEEE Transactions on Information Theory*, for example, has devoted a special issue (March 1992, Part II) with 31 papers to this topic and Academic Press has begun a series of books devoted to wavelets. Daubechies, who perhaps has done more fundamental work on wavelets than any other researcher, was awarded a MacArthur prize this year. Considerable popular interest also exists ("Catch a Wave", 1992; Kolata, 1991; Wallich, 1991 and Carey, 1992). Actual applications with wavelets are either in research or soon to appear. The digital compact cassette (DCC), recently introduced by Phillips (which is said to have CD sound quality) uses a wavelet–like method called subband coding. Wavelets may also allow magnetic resonance images to be generated instantaneously, thereby removing the lengthy exposure time that limits this very useful imaging technique (Healy and Weaver, 1992). A startup company devoted to wavelets is even in business.

Very briefly, the discrete wavelet transform is a two dimensional orthogonal decomposition of a time series that is well suited, and is in fact designed, to detect abrupt changes and fleeting phenomena. The orthogonality guarantees a unique decomposition, and the two dimensionality allows the series to be studied at different scales. One very important characteristic of the discrete wavelet transform used here is that its basis functions have compact support, i.e., are non–zero for only a limited range. Thus, they are able to pick up unique phenomena in the data. By contrast, the basis function for Fourier transforms, sines and cosines, have infinite support and are less well suited for detecting such phenomena in the data. In another contrast to Fourier methods, the discrete wavelet transform can be used to study nonstationary data. The implications for macro data are obvious and are demonstrated below.

This paper first describes the discrete wavelet transform and then studies their use in analyzing macroeconomic data. These initial results are quite favorable and more work is in order.

D. A. Belsley (ed.), Computational Techniques for Econometrics and Economic Analysis, 137–149.

2. DILATION EQUATIONS, MOTHER FUNCTIONS AND WAVELETS

a. References on Wavelets

In this and the next section wavelets are described in a very intuitive fashion. More rigorous descriptions can be found elsewhere. Press (1992) has a particularly good description of the wavelet decomposition algorithm and contains the Fortran code to perform it. Strang (1989) provides a more mathematical treatment. Rioul and Vetterli (1991) describe wavelets from a signal processing perspective. The ultimate authority is Daubechies (1988).

b. Mother Functions and Dilation Equations

Wavelets begin with a solution to a dilation equation, which generates so–called mother functions (also called scaling functions). These mother functions are intimately related to wavelets and, in fact, are used in the discrete wavelet transform. Equation (1) shows the specific dilation equation used in this literature:

$$\phi(x) = \sum_{k=0}^{m} c_k \phi(2x - k)\,. \tag{1}$$

The $m + 1$ values of c_k determine the mother function ϕ. This equation is recursive: the sum of modified forms of ϕ equals ϕ itself. Viewed in graphical terms, it can be seen that the 2 in the right hand argument of (1) shrinks ϕ horizontally, the k shifts ϕ horizontally, and the c_k shrink or expand ϕ vertically.

As a simple initial example, consider the case with $m = 1$ and $c_0 = c_1 = 1$ so that (1) becomes

$$\phi(x) = \phi(2x) + \phi(2x - 1)\,. \tag{2}$$

The solution to (1) is the box function, which has value 1 over $[0, 1]$ and is zero elsewhere. The coefficient 2 in the argument of the first term on the right hand side of (2) shrinks ϕ horizontally by a factor of 2. Without the shift given in the second term, the box extends from 0 to $\frac{1}{2}$. The second term on the right hand side of (2), however, shifts the narrowed ϕ right by $\frac{1}{2}$, thus completing the solution. This is illustrated in Figures 1 and 2, which show the left and right hand sides of (2), respectively.

A very interesting and useful family of orthogonal mother functions was discovered by Daubechies (1988). Members of this family are both orthogonal and have compact support, and are categorized by their ability to form polynomials of different orders. Each member of this family, when taken as a linear combination with itself, generates a given order polynomial. The first member is the box function; when taken in linear combination with itself, it forms a constant value (i.e. a polynomial of degree zero). The second member, which has four c_k coefficients, is called D_4 by Strang (by this notation, D_2 is the box function). D_4, taken in linear combination with itself, forms a line of arbitrary constant slope (i.e., a polynomial of degree one). Its coefficients, in equation (3), are more complicated than those of the box function.

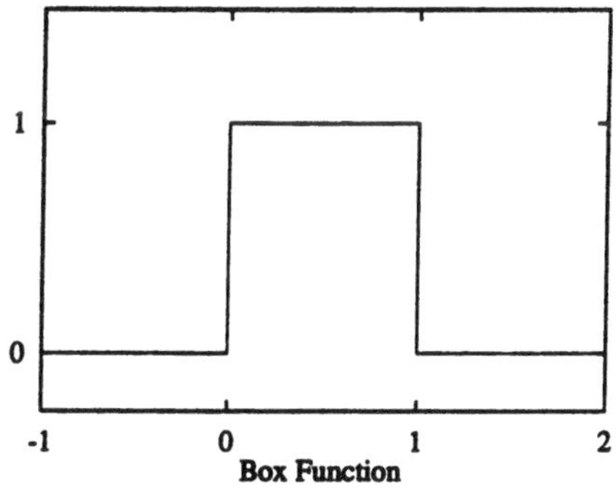

Fig. 1.

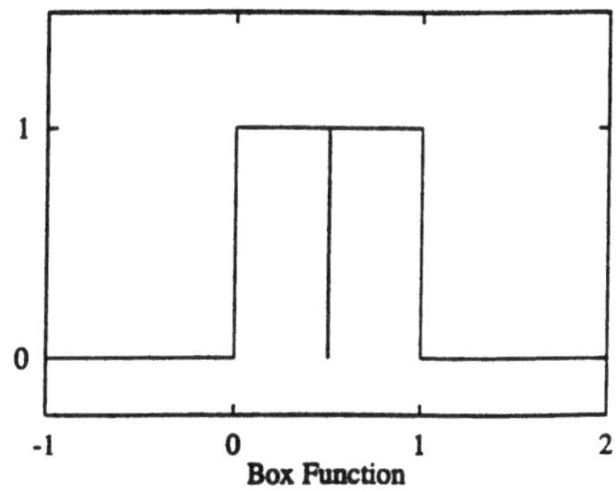

Fig. 2.

$$\begin{aligned} c_0 &= \tfrac{1}{4}(1+\sqrt{3}) \qquad & c_2 &= \tfrac{1}{4}(3+\sqrt{3}) \\ c_1 &= \tfrac{1}{4}(3+\sqrt{3}) \qquad & c_3 &= \tfrac{1}{4}(3-\sqrt{3})\,. \end{aligned} \tag{3}$$

Higher order members of this family generate higher order polynomials, and, as can be seen in going from D_2 to D_4, with each increase in the order of the replicated polynomial, the number of c_k's increases by two.

Figure 3 shows the D_4 mother function. Its asymmetry is striking, and while it is everywhere continuous, it is not everywhere differentiable. Specifically, it is not left differentiable at so–called dyadic points: $k/(2^j)$ for k and j integers (Pollen, 1992). This bizarre shape arises from the stringent requirements of local support and orthogonality. A linear combination of D_4 mother functions forms a polynomial of degree one because the cusps cancel one another.

An important characteristic of mother functions and wavelets is that many are constructed recursively; they cannot be written out in closed form like most more familiar functions. For instance, Figure 3 was generated by using equation (1) recursively with c_k as given in (3) and initial values for $\phi(1)$ and $\phi(2)$. Equation (1) first yields $\phi(\frac{1}{2})$, $\phi(1\frac{1}{2})$ and $\phi(2\frac{1}{2})$. Next, values at $\frac{1}{4}$ intervals can be constructed, and so on. The function is non–zero only on $[0,3]$.

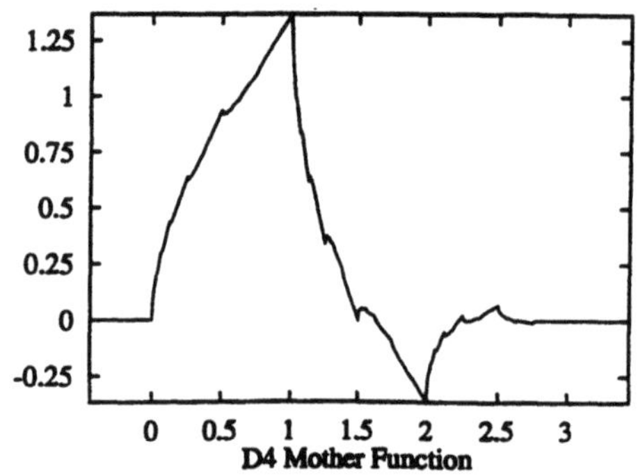

Fig. 3.

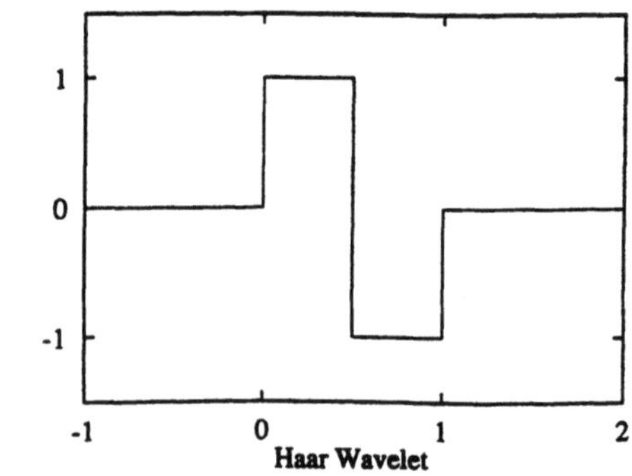

Fig. 4.

c. *Wavelets*

A wavelet is described by

$$W(x) = \sum_{k=0}^{m} (-1)^k c_{1-k} \phi(2x - k) \,. \tag{4}$$

$W(x)$ clearly bears a very close connection to (1); a wavelet is basically a rearranged mother function. Figure 4 shows the Haar wavelet whose mother function is the box function, and W_4 is likewise similar to its mother function, D_4.

3. THE DISCRETE WAVELET DECOMPOSITION

The discrete wavelet decomposition is best understood in comparison to the discrete Fourier decomposition. Equation (5) shows the discrete orthonormal cosine Fourier decomposition for a series of length 4 (this short length shows its essential elements).

$$\begin{bmatrix} x(1) \\ x(2) \\ x(3) \\ x(4) \end{bmatrix} = \frac{a_1}{\sqrt{2}} \begin{bmatrix} \frac{1}{2} \\ \cos(\pi/8) \\ \cos(2\pi/8) \\ \cos(3\pi/8) \end{bmatrix} + \frac{a_2}{\sqrt{2}} \begin{bmatrix} \frac{1}{2} \\ \cos(3\pi/8) \\ \cos(6\pi/8) \\ \cos(9\pi/8) \end{bmatrix}$$

$$+ \frac{a_3}{\sqrt{2}} \begin{bmatrix} \frac{1}{2} \\ \cos(5\pi/8) \\ \cos(10\pi/8) \\ \cos(15\pi/8) \end{bmatrix} + \frac{a_4}{\sqrt{2}} \begin{bmatrix} \frac{1}{2} \\ \cos(7\pi/8) \\ \cos(14\pi/8) \\ \cos(21\pi/8) \end{bmatrix} \tag{5}$$

The basis functions for this decomposition consist of the four vectors on the right hand side. The coefficients $a_1, \ldots, a_4$ are the discrete cosine transform of the vector x. This equation clearly shows how a series can be decomposed into a sum of sinusoids that vary in frequency. But it is important to note that these basis functions have infinite support; that is, no elements of the basis vectors are zero. As a result, Fourier type decompositions have difficulty picking out changes in x that are permanent or fleeting; they are best for series that are stationary.

Equation (6) shows a wavelet decomposition of a time series of length 8 with the Haar wavelet (as shown below, this longer length better illustrates the decomposition's key features since it contains a variety of different sized wavelets).

$$\begin{bmatrix} x(1) \\ x(2) \\ x(3) \\ x(4) \\ x(5) \\ x(6) \\ x(7) \\ x(8) \end{bmatrix} = \frac{b_{01}}{2} \begin{bmatrix} 1 \\ 1 \\ 1 \\ 1 \\ 0 \\ 0 \\ 0 \\ 0 \end{bmatrix} + \frac{b_{02}}{2} \begin{bmatrix} 0 \\ 0 \\ 0 \\ 0 \\ 1 \\ 1 \\ 1 \\ 1 \end{bmatrix} + \frac{b_{11}}{2} \begin{bmatrix} 1 \\ 1 \\ -1 \\ -1 \\ 0 \\ 0 \\ 0 \\ 0 \end{bmatrix} + \frac{b_{12}}{2} \begin{bmatrix} 0 \\ 0 \\ 0 \\ 0 \\ 1 \\ 1 \\ -1 \\ -1 \end{bmatrix}$$

$$+ \frac{b_{21}}{\sqrt{2}} \begin{bmatrix} 1 \\ -1 \\ 0 \\ 0 \\ 0 \\ 0 \\ 0 \\ 0 \end{bmatrix} + \frac{b_{22}}{\sqrt{2}} \begin{bmatrix} 0 \\ 0 \\ 1 \\ -1 \\ 0 \\ 0 \\ 0 \\ 0 \end{bmatrix} + \frac{b_{23}}{\sqrt{2}} \begin{bmatrix} 0 \\ 0 \\ 0 \\ 0 \\ 1 \\ -1 \\ 0 \\ 0 \end{bmatrix} + \frac{b_{24}}{\sqrt{2}} \begin{bmatrix} 0 \\ 0 \\ 0 \\ 0 \\ 0 \\ 0 \\ 1 \\ -1 \end{bmatrix}. \tag{6}$$

The first two vectors on the right hand side are the mother function for the Haar wavelet, which of course is the box function (the terms in the coefficients' denominators are part of the wavelet or mother function and make them orthonormal). The other six vectors are the actual Haar wavelet at two scales and locations. The first two of these are on a larger scale than the second four. The discrete wavelet transform consists of the b_{ij} coefficients. The two subscripts demonstrate the two-dimensional nature of the decomposition, one of its key features. These coefficients are categorized by $i = 0$, 1 and 2. For $i = 0$, the basis is the mother function. For $i = 1$, the basis is the largest wavelet, and for $i = 2$, the smallest wavelet. In general, a wavelet decomposition first contains two coefficients for the mother function. Then, defining the length of a wavelet by its number of non–zero elements, coefficients for wavelets cover half the series' length, then a quarter, an eighth, and so on, until the smallest, which have two non–zero elements. As the coverage of each wavelet shrinks, the number of wavelets of that size doubles. In (6), this can be seen when moving from the third and fourth vectors to the fifth through eighth ones. Overall, this means that the series to be decomposed must have a length that is 2 to a power (i.e. 2, 4, 8, 16, 32, ...). Unfortunately, this imposes some unwanted restrictions on the length of the series one can study. For example, it does not appear that the process can be tricked by padding the series with additional data, since any additional data would influence the decomposition.

Most importantly, each wavelet contains a considerable number of zero elements (the so–called compact support). Thus, the discrete wavelet transform has the potential to "pick up" unique phenomena in the data.

Wavelets of different lengths refer to properties of the data at different frequencies, though the preferred term in this literature is scale. The longest wavelets contain elements of the data at low frequency and large scale, while the smallest wavelets embody high frequencies and small scales.

To further illustrate wavelet decompositions and lay the groundwork for this display in this paper, consider the following actual decomposition:

$$\begin{bmatrix}1\\2\\1\\3\\10\\0\\2\\1\end{bmatrix} = \frac{3.5}{2}\begin{bmatrix}1\\1\\1\\1\\0\\0\\0\\0\end{bmatrix} + \frac{6.5}{2}\begin{bmatrix}0\\0\\0\\0\\1\\1\\1\\1\end{bmatrix} - \frac{.5}{2}\begin{bmatrix}1\\1\\-1\\-1\\0\\0\\0\\0\end{bmatrix} + \frac{3.5}{2}\begin{bmatrix}0\\0\\0\\0\\1\\1\\-1\\-1\end{bmatrix}$$

$$-\frac{\sqrt{2}/2}{\sqrt{2}}\begin{bmatrix}1\\-1\\0\\0\\0\\0\\0\\0\end{bmatrix} - \frac{\sqrt{2}}{\sqrt{2}}\begin{bmatrix}0\\0\\1\\-1\\0\\0\\0\\0\end{bmatrix} + \frac{5\sqrt{2}}{\sqrt{2}}\begin{bmatrix}0\\0\\0\\0\\1\\-1\\0\\0\end{bmatrix} + \frac{\sqrt{2}/2}{\sqrt{2}}\begin{bmatrix}0\\0\\0\\0\\0\\0\\1\\-1\end{bmatrix}. \quad (7)$$

Note how the original series on the left hand side has a "spike" – the fifth term is much larger than the others – and how the coefficient on the third of the smallest wavelets (the seventh vector on the right hand side), located at the site of the spike, reveals this phenomenon with its relatively large coefficient.

Figure 5 displays the discrete wavelet transform of (7). The bottom panel shows the actual time series arranged horizontally with the smallest values on the bottom.

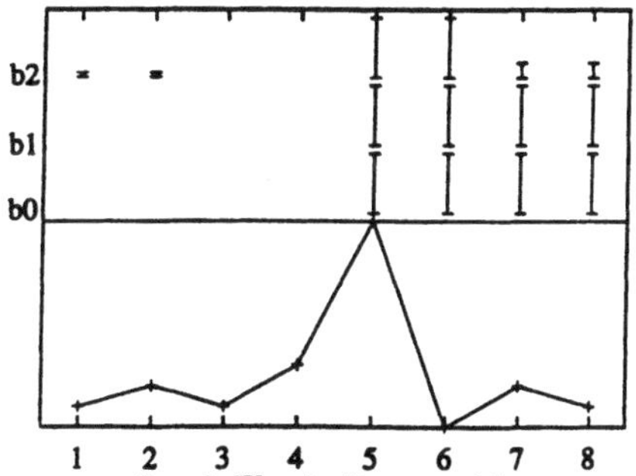

Sample Wavelet Decomposition

Fig. 5.

Note the spike in the data at the fifth observation. The top panel shows the wavelets' coefficients. The convention followed in their display uses the relative size of the coefficeints within each category of wavelet. In the top panel, the b_{0j} coefficients are on the bottom row, the b_{1j} coefficients are in the middle row, and the b_{2j} coefficients are on the top (the b_0, b_1 and b_2 terms on the left hand side of the figure denote the rows). The coefficients in each of these three rows are arranged in terms of their relative size, so the largest coefficient value "fills" its location in that row while the smallest leaves it empty. Intermediate values fill the height of the row proportionally. In (7), b_{01} is associated with the first half of the time series, and b_{02} is associated with the second half, given the location of the mother functions in the first two vectors. With $b_{01} = 3.5$ and $b_{02} = 6.5$, the first half of the bottom row, associated with the smaller value 3.5, is empty, while the space to the right is full. Likewise, the middle row, that for b_{1j}, has only two parts, so the left half, associated with the smaller b_{1j} coefficient, is empty while the other half is full. Note that in both of these rows, the wavelet coefficients cover a length of four, thus accounting for the four lines. In the top row, that for b_{2j}, there are four parts. Again the largest coefficients fill the panel height; the smallest are empty and the intermediate values are so noted. In the displays that follow, the first two rows of the wavelet coefficients are frequently dropped since they convey little information (one part being invariably the largest and the other the smallest) and take up valuable space.

The algorithm that generates the discrete wavelet transform is relatively simple and quite fast. An intuitive description of the algorithm is given in Press (1992). Briefly, a matrix is first formed with the c_k's oriented roughly along the diagonal. The data vector is multiplied by this matrix, and the resulting vector is sorted. Another multiplication then takes place with half of the sorted vector and the matrix of the c_k's. This process of multiplication, sorting, and discarding half the vector continues until the sorted vector contains two elements. The vector formed from the discarded elements contains the discrete wavelet transform coefficients.

The algorithm's operation count is proportional to n, the length of the original data and is therefore faster than the fast Fourier transform, whose operation count is proportional to $n \cdot \log_2(n)$. Thus, there are no computational constraints to the discrete wavelet transform for most economic data and certainly none for macro data.

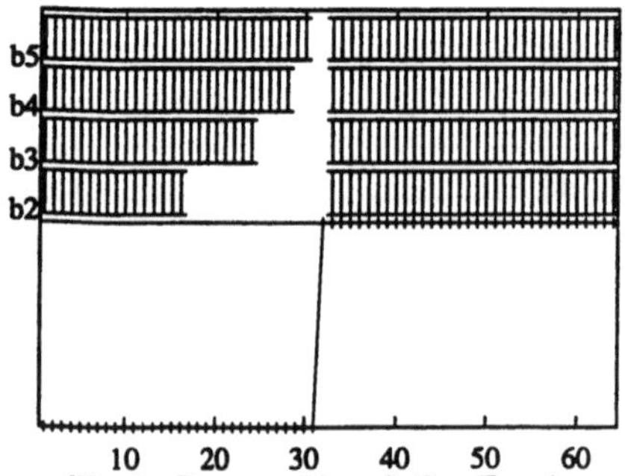

Fig. 6.

4. THE DISCRETE WAVELET TRANSFORM ON GENERATED TIME SERIES

To better understand how the discrete wavelet transform decomposes data, we use it here on several pathological time series. Figure 6 shows the graphics associated with a discrete wavelet transform of a step series of length 64 (recall that the series to be studied must have length of 2 to an integer power). It has a value of 1 prior to observation 31 and jumps to 1.1 thereafter. The Haar wavelet seems to do a better job than other wavelets on macro series (as described below), and so is employed in this section. As can be seen, there is an abrupt change in all categories of the wavelet coefficients at the point where the time series jumps. Note the bottom row of the wavelet panel contains 4 wavelet coefficients (the first, third, and fourth have the same size), each covering a length of 16, the next has 8 of length 8, the next 16 of length 4, and the final 32 of length 2. The wavelet coefficients with only two entries are not shown. In every row, all coefficients but one are the same size (this accounts for the all or nothing height in the blocks). The change at the jump in the original series is seen with all coefficients, even though they cover vastly different scales and provide different resolutions for marking the break.

Figure 7 shows the decomposition of a sine wave. Once again, increased resolution and detail are seen as one moves up the coefficient panel. This figure is most instructive when compared with Figure 8, whose frequency is the same as that in Figure 8, and then increases by 20% at the midpoint. While the change in frequency is barely noticeable when comparing the two time series, note the very different low frequency (long scale) behavior with the two sets of wavelet coefficients in the two figures. Further, a comparison of the higher frequency wavelet coefficients shows differences as well. While not as striking as Figure 6, it nonetheless shows the change in the series at its midpoint.

Figure 9 shows a series whose slope increases by 20% at its midpoint. While the change is barely visible to the eye, it is quite obvious by looking at the wavelet coefficients panel that a major change has taken place in the actual location of the change in the slope.

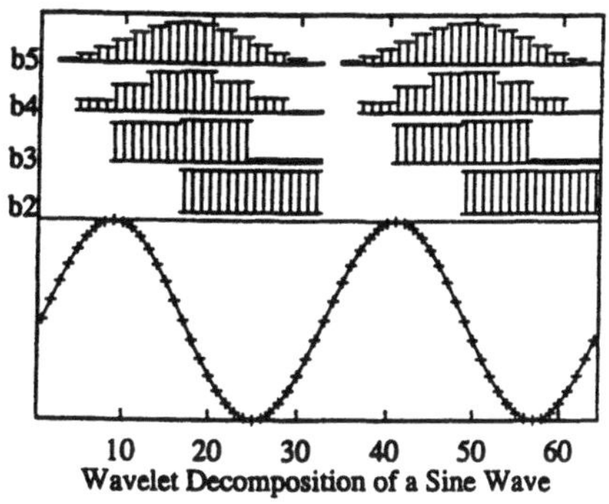

Fig. 7.

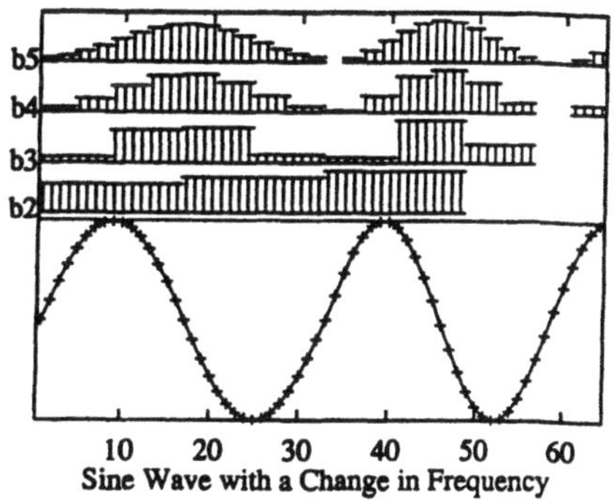

Fig. 8.

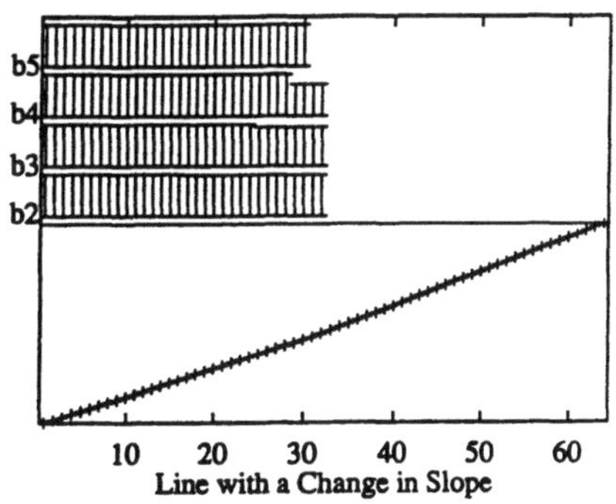

Fig. 9.

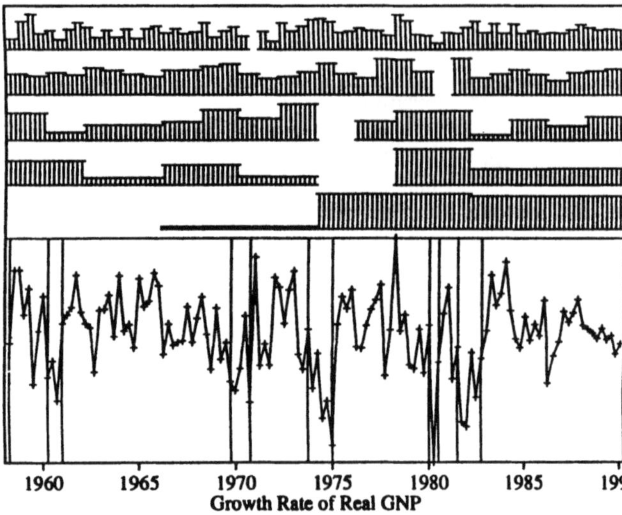

Fig. 10.

5. THE DISCRETE WAVELET TRANSFORM ON THREE MACRO SERIES

This section is quite speculative since the work on wavelets in macro is very preliminary. However, it does suggest some interesting uses for wavelets in economics. The Haar wavelet is used for all decompositions in this section since it proved to generate the most interpretable decompositions. The D_4 wavelet yielded decompositions of macro data that did not correspond to known features in the data, such as business cycles and secular changes. Other wavelets were not investigated.

Figure 10 shows the discrete wavelet decomposition of the quarterly growth rate of real GNP for the period from 1958II to 1990I (128 observations). Business cycle turning points are denoted by vertical lines in the bottom panel (dates determined by the NBER). In turn we see the end of the 1957–1958 recession, the 1960–1961 recession, the 1969–1970 recession, the 1973–1975 recession, the 1980 recession, and finally, the 1981–1982 recession. This figure is not particularly revealing; the major revelation is the low and mid–frequency change in 1974, which appears to capture the well–known slowdown in the economy's growth rate. The high frequency components cannot be identified with any particular phenomena. Rather surprisingly, the wavelet decomposition reveals only long term phenomena on real GNP's growth rate.

Figure 11 shows the results for the level of the log of real GNP, and it conveys more information than the previous case. In every recession but the first, which is only partly included in the data, the highest frequency wavelet coefficients are quite large. With the single exception of a period in 1959 (a downturn, but not a recession), they are almost never as large in recessions, and so these large high frequency coefficients identify recessions. These results hold up even better in the wavelet coefficients with the next lowest frequency, though naturally with less resolution. At the low frequencies, one sees significant changes in 1966I to 1966II and 1974I to 1974II. The

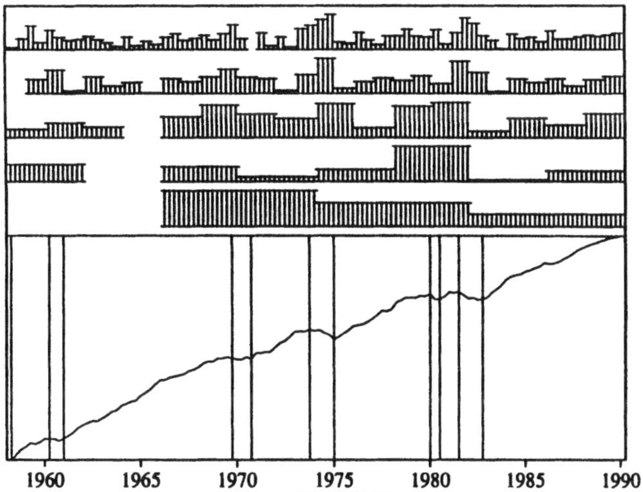

Fig. 11.

later seems to signify the change in the growth rate in the early 1970's. Overall, this figure is rather impressive since it uses unfiltered data and it picks up both low and high frequency items of interest.

Figure 12 examines the change in nominal nonfarm business sector compensation per hour (Citibase series LBCPU) over the same dates as the real GNP data. The first thing we see is how the spikes in the data are picked up by the high frequency coefficients. We also note that the two lowest frequency components pick up breaks after 1966I, 1974I and 1982I with some evidence for a break after 1970I in the next to lowest frequency coefficient. Balke (1991) studied this series for evidence of level shifts using a modified version of Tsay's method for detecting outliers (1988). He identified level shifts in 1968I, 1972IV, and 1982II. Given the low resolution of low frequency wavelets, the match between these two very different methodologies is remarkable.

Figure 13 analyzes a synthetic series. The data are generated from an ARIMA model that Blanchard and Fischer (1989) fit to GNP growth. Thus, it is designed to replicate the time series in Figure 10. While the two time series appear to have the same characteristics to the eye, notice the relatively greater variation in the high frequency components in Figure 13 than in Figure 10 (this is somewhat hard to see given the different sizes of the figures). This suggests that the discrete wavelet transform may be useful as a diagnostic tool for fitting models or as a method for detecting data characteristics ignored by models.

6. CONCLUSION

This paper illustrated intuitively the discrete wavelet transform and then went on to explore its usefulness for macroeconomics. Although very preliminary, initial

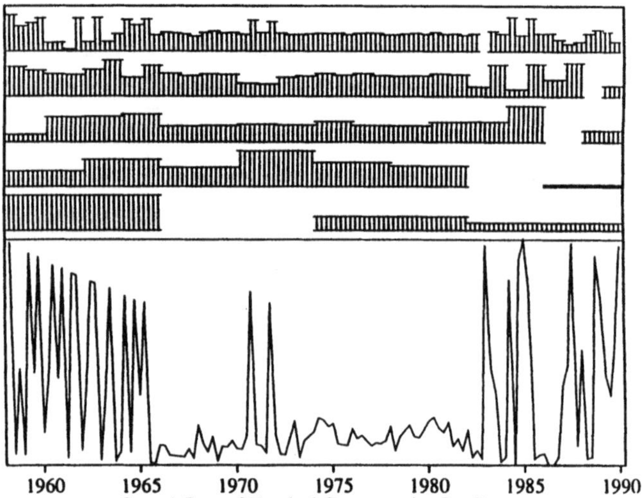

Fig. 12.

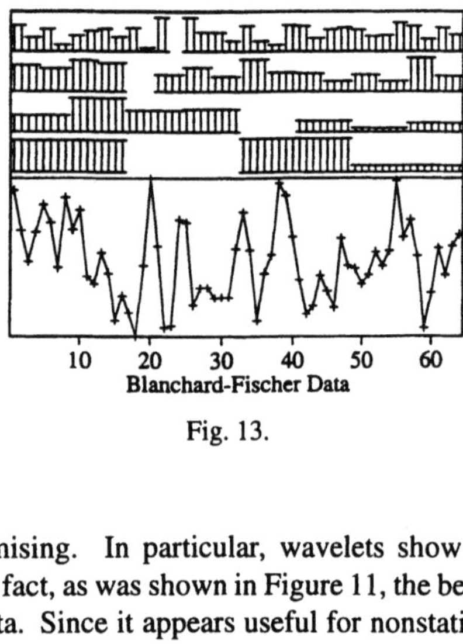

Fig. 13.

results are quite promising. In particular, wavelets show potential in analyzing nonstationary data. In fact, as was shown in Figure 11, the best results were obtained with nonstationary data. Since it appears useful for nonstationary data, the method also shows potential for examining data simultaneously at low and high frequencies, so there is no particular short–run/long–run distinction. This is of particular macro interest given that real business cycle theorists reject the conventional macro short–run/long–run distinction. Finally, unlike Fourier methods, wavelet analysis locates phenomena of interest in a time series. For instance, Figure 11 showed that recessions and secular changes could easily be identified.

Further possible uses of the discrete wavelet decomposition include smoothing macro time series or isolating short–term phenomena. Either of these could be accomplished by reconstructing the data with only the large-scale or small-scale wavelet coefficients, respectively. The use of the wavelet decomposition as a diagnostic tool for model building was demonstrated with data generated from the Blanchard–Fischer equation, which was shown in Figure 13 to have a different decomposition than actual data. Finally, it might be used not only as a diagnostic tool for statistical and econometric models, but also for data generated from macroeconomic models.

ACKNOWLEDGEMENTS

I would like to thank, without implicating, Nathan Balke, Tony Smith, David Belsley and the participants of the session titled "Computational Elements in Econometrics and Statistics II" at the 14th Annual Congress of the Society for Economic Dynamics and Control in Montreal in June, 1992.

REFERENCES

Balke, Nathan S. "Detecting Level Shifts in Time Series: Misspecfication and a Proposed Solution." Richard B. Johnson Center for Economic Studies Working Paper #9121. Department of Economics, Southern Methodist University. 1991.

Blanchard, Olivier Jean and Stanley Fischer. *Lectures on Macroeconomics.* Cambridge MA: MIT Press, 1989.

Carey, John. "'Wavelets' Are Causing Ripples Everywhere." *Business Week.* Febuary 3, 1992: 74–5.

"Catch a Wave." *The Economist.* 323 no. 7754 (1992): 86.

Daubechies, Ingrid. "Orthonormal Bases of Compactly Supported Wavelets." *Communications in Pure and Applied Mathematics* 41 (1988): 909–96.

Healy, Dennis M. Jr. and John B. Weaver. "Two Applications of Wavelet Transforms in Magnetic Resonance Imaging." *IEEE Transactions on Information Theory.* 38 (1992): 860–880.

Kolata, Gina. "New Technique Stores Images More Efficiently." *New York Times.* November 12, 1991: B5+.

Pollen, David. "Daubechies' Scaling Function on [0,3]." *Wavelelts: A Tutorial in Theory and Applications.* Ed. Charles K. Chui. San Diego, CA: Academic Press, 1992.

Press, William H., Saul A. Teukolsy, William T. Vetterling and Brian P. Flannery. *Numerical Recipes in Fortran, The Art of Scientific Computing, second edition.* New York: Cambridge Universtiy Press, 1992.

Rioul, Olivier and Martin Vetterli. "Wavelets and Signal Processing." *IEEE Signal Processing Magazine.* October 1991: 14–38.

Strang, Gilbert. "Wavelets and Dilation Equations: A Brief Introduction." *SIAM Review* 31 (1989): 616–27.

Tsay, R. S. "Outliers, Level Shifts, and Variance Change in Time Series." *Journal of Forecasting* 7 (1988): 1–20.

Wallich, Paul. "Wavelet Theory: An Analysis Technique that is Creating Ripples." *Scientific American* January 1991: 34–35.

C.R. BIRCHENHALL*

MatClass: A Matrix Class for C++

ABSTRACT. The MatClass project is an experiment in the use of object-oriented methods in numerical methods using C++. MatClass is a family of numerical classes for C++ that are freely available. MatClass combined with a C++ compiler gives the user a compiled matrix language together with a set of numerical and statistical classes based on the key matrix decompositions e.g. LU, Cholesky, Householder QR, and SVD. It is argued that C++ with classes such as MatClass offer a valuable third line of development complementing the current standards of Fortran and GAUSS. While object-oriented numerical programmming is not revolutionary, it is a significant new development. This paper aims to give a brief and superficial overview of the current state of MatClass and to announce the availability of Version 1.0d.

1. INTRODUCTION

Given that computing is an increasingly important aspect of our working environment, the profession should be concerned to promote good computing practice amongst its members, just as it promotes good practice in the formulation of models and the statistical testing of hypotheses. This is not the place to pursue this theme in depth; suffice it to say that my aim in providing an open and free programming tool is to contribute to the development of an embryonic computational economics.

Essentially, software development has been seen as a commercial activity rather than a fundamental contribution to the discipline. This is not to suggest the authors of the standard packages are not appreciated by the profession; on the contrary every applied economist will greatly value the services provided by the author of his chosen system. The problem is that the reliance on commercial incentives invariably means that source code is kept secret, and so software is removed from peer review and essentially closed to third party improvement and extension. No doubt one of the

* This document is partly derived from an earlier piece written jointly by myself and Jarlath Trainor. Many of the strengthens of the current document reflect Jarlath's contributions, but he has no responsibility for the errors and weaknesses that have undoubtedly arisen from my rewrite of this document or MatClass. Many thanks for Jarlath's assistance on this project. My thanks also to the Department of Econometrics & Social Statistics for funding Jarlath's time in the department. Anyone who is working with Unix workstations will know the value of having a Unix wizard at hand; in my own case I am greatly indebted to Owen LeBlanc from the Manchester Computing Centre. Finally my thanks to Ericq Horler and David Belsley for their careful reading of an earlier version of this document and the removal of many errors. While their efforts has greatly improved this work they cannot be held responsible for the remaining inadequancies.

D. A. Belsley (ed.), Computational Techniques for Econometrics and Economic Analysis, 151–172.

reasons for this has been the perceived complexity of software development and the scarcity of requisite expertise. The success of GAUSS suggests that with appropriate tools a significant number of economists will turn their hand to programming rather than rely solely on the menu provided by the packages. If there is to be any prospect of developing a subculture of economic code, there needs to be an appropriate set of tools and standards. Recently there has been a good deal of interest in using object-oriented programming systems (OOPS) in reducing the difficulties of software development; in particular there has been a flourish of interest in the C++ language. The central aim of the MatClass project is to go some way in evaluating the potential for an object-oriented approach to econometric computation using C++.

2. FORTRAN, GAUSS AND C++

Currently the choice of standards for programming in economics seems to be between Fortran and GAUSS. MatClass aims at contributing to a third front, namely an object-oriented approach based on C++. To clarify the potential role of C++, we first take a quick look at Fortran and GAUSS.

Fortran's long history has led to a vast amount of code being available in various forms, including free and open libraries and numerous published algorithms. Most numerical analysts still see Fortran as their *lingua franca*. From the perspective of "openness" Fortran ranks highly. Fine examples of freely available code are LINPACK[4] and *Numerical Recipes*[12]. On the positive side Fortran offers *flexibility* and *portability*. Fortran's flexibility arises from the compiled nature of the runtime programs; that is to say the source code is translated into machine code for the target hardware. The significance of this is that new algorithms can replace or extend existing library routines without loss of efficiency. In contrast, GAUSS, which from this perspective is an interpreter, makes 'third party' code suffer from runtime inefficiences. If code is truly to be open then there should be few impediments to correcting, improving, or extending the code. Fortran's portability is based on the popularity of the language, at least in science and engineering, which in turn means most university systems have a Fortran compiler. Equally important is the development of standards, in particular the standardisation of the language itself.

On the negative side, Fortran, at least prior to Fortran 90, suffered from lack of support for aggregate structures and various control structures. In contrast to GAUSS, Fortran 77 had no concept of a matrix, though Fortran 90 remedies this matter. In contrast to Pascal or C, Fortran 77 does not support control structures such as "while" or "case" statements — though some see the latter as a strength ([13] p. 13). Fortran 90 remedies these matters.[1] In contrast to C++ and other object-oriented languages, Fortran 77 and 90 lack the concept of inheritance, the significance of which will be discussed in Section 4.

[1] I cannot recall who it was who suggested that the old Algol-Fortran argument was now settled in Algol's favour, but it seems to be a fair comment on Fortran's belated acceptance that there is merit in structures and the "new" controls.

Turning to GAUSS, we can readily understand its popularity by considering its matrix syntax, which is supplied in a relatively simple interpretive environment with efficient intrinsics. A matrix syntax leads to significant increases in programmer productivity, particularly in econometrics. As an interpreter with extensive and efficient intrinsics, GAUSS offers the user a "super calculator", where many statistical calculations can be executed with relative ease. Indeed I would suggest that GAUSS is to the econometrician what a spreadsheet is to an accountant — without wishing to suggest that GAUSS programs are as intractable as Lotus macros. Unlike Fortran and other compiled languages, GAUSS does not suffer from compiler and linker lags. This latter feature is particularly important when working on older PCs with relatively low computational power and slow disks. In contrast to Fortran, one of the main drawbacks of GAUSS is the impediment it places on code replacement and extension. A Fortran programmer can mix and match a library of routines without loss of runtime efficiency. Low-level GAUSS code, i.e. code that involves layered loops addressing individual elements of a matrix, is relatively inefficient, even when compared with GAUSS' own intrinsic and fully compiled code, a situation that arises from being an interpreter.

A major difficulty with the original GAUSS was its lack of portability; it is not possible to port its Intel based assembler to the RISC based Unix workstations. There is currently an expectation, however, that GAUSS will become available for SUN workstations, and because this rewrite is based on C, GAUSS may become available to other workstations. But the underlying code is not in the public domain, and the porting to other systems will depend on the ability and willingness of the supplier. A clear advantage to public domain code is that any interested party can initiate ports to their chosen system, and portability is important if economic researchers are to exploit the workstation model of computing. While the new 486 PCs are powerful they fail to match the standards set by systems such as SUN's SPARC and HP's Series 700 workstations. This is not simply a matter of raw computing power — though that is significant, see Section 9 — it is a matter of the stability and capability of operating systems. Indeed I have found that my own move to such a machine has led to significant increases in productivity, and I now find it difficult to return to a 386 PC with DOS. Furthermore, I am confident that the cost of workstations will continue to decline and they will soon be feasible for most researchers — be it a 486, or 586 with Windows/NT[2], or RISC machines with Unix. What is also clear is that there will be several contenders in this market, namely SUN, HP, IBM, DEC and various Intel based kits, and thus portability of software will be an issue.

It has to be said that there are some aspects of the GAUSS language that are less than desirable. In particular, the restricted range of data types (matrix and string), together with the limited means for extending the type system (arrays), does not encourage structured programming. While the matrix type is an improvement on Fortran's restriction to scalar types and arrays, some general form of record or

[2] MS-Windows suffers from poor memory management and instability due to its dependance on DOS. Windows/NT or even OS/2 v2 hold more promise for sustained development of Intel based systems.

structure is extremely powerful and sadly missing from GAUSS. Needless to say GAUSS offers no form of inheritance between types.

One minor concern is the long term support of GAUSS. Unlike Fortran with its multiple sources and open libraries, GAUSS comes from a single source, and the underlying code is closed. As far as I know there is no secondary source for GAUSS interpreters, and this could be a major problem if this single source should fail. This is not to suggest GAUSS should have no long term role; indeed it will undoubtedly remain a major force in code development but I would be concerned to become solely dependent on the current system.

It can be noted that most of the above comments on GAUSS apply to MATLAB, although it is already available on most systems. While few econometricians are using MATLAB, it is a system offering a similar set of basic facilities, although it lacks some of the statistical intrinsics and does not have the various add-ons. Coming from one of the authors of LINPACK, it has found favour among numerical analysts[7]. It has been suggested that MATLAB not only facilitates the learning of numerical methods, but it can also be used for prototyping new code before being translated into Fortran. A similar role could be, and possibly is being, played by GAUSS.

So where does C++ stand in relation to these existing standards? What makes C++ interesting is its promise of combining object-oriented methods with runtime efficiency. But C++ is currently more promise than fact and will not provide a serious challenge to GAUSS or Fortran until there appropriate *class* libraries on which users can build. C++ uses the concept of a **class** to allow object-oriented extensions to the C language, i.e. developers can use the class system to extend the object types supported by the compiler. MatClass is one of several classes that defines a matrix type and gives the C++ user a matrix syntax and other facilities similar to those offered by GAUSS or MATLAB.[3] But that is not all, for the set of possible data types are limited only by the programmer's imagination. MatClass uses higher level classes to embody higher level functionality. For example, it offers a family of classes to embody ordinary least-squares models. These OLS classes give the C++ programmer a "package" that hides the storage and computational details underlying the least-squares procedures. At the other extreme, MatClass has low-level classes for debugging and error management. For example there is a generic class, matObject, with error management capabilities *whose features are inherited by all other objects*. These few sentences cannot give the reader a complete picture of object-oriented programming, but we hope we have conveyed the image of a flexible and extensible system for which MatClass is but a simple beginning.

In comparing the C++ and MatClass combination with GAUSS and Fortran, note that MatClass acts as a complement to a C++ compiler which links the user program with the MatClass library before execution. MatClass based programs, then, are compiled to machine code, having the advantage of efficient low-level routines but the disadvantage of compiler and linker lags. Unlike Fortran, C++ is a young

[3] A commercial based class that offers similar, if not greater, facilities is M++ from Dyad Software.

language, and there are few numerical libraries available. Fortunately, C++ is a superset of C, and thus existing C libraries, in particular *Numerical Recipes in C* [13], are readily adapted for use in C++. MatClass aims to make some contribution in this direction. In its favour, when compared to Fortran 77, MatClass gives the C++ programmer a matrix syntax and an object-oriented environment. Apart from the issue of compiler versus interpreter, the comparison of MatClass and GAUSS focuses on C++'s support for, and MatClass's exploitation of, object-oriented programming. A fuller discussion of OOPS will be offered below. Here we simply emphasize that use of C++ classes gives the developer a highly structured language with strong type checking, and note that most compilers come with source level debuggers that, together with the debugging features of MatClass, make the debugging of larger projects *relatively* straightforward.

Some might wonder whether a highly structured approach to numerical computing with layers of types and classes might not suffer from runtime inefficiences. It will be demonstrated that the combination of MatClass with a good quality compiler produces code that is efficient, particularly for low-level tasks. Furthermore the code is immediately portable to RISC based Unix — for example, timings will show a HP 720 workstation running up to 10 times faster than a 33Mhz 486 PC.

All in all, then the combination of (a) having the source code in the public domain, (b) the compiled nature of user written routines and (c) C++'s support for OOPS, should make MatClass a good candidate to act as a foundation for an "open sytem" of software. While C++, with appropriate classes, will not replace Fortran or GAUSS, it promises to be a valuable third leg for developers.

3. A QUICK LOOK AT C++ AND MATCLASS

An extensive evaluation of C++ cannot be given here, but a few words are in order. C++ is a wide spectrum extension of C that allows the programmer to combine low-level and object-oriented styles in the same program. In part C++ can be viewed as a better and extended version of the C programming language. From this perspective C++ is seen to inherit from C its *efficiency, flexibility* and *portability*. As a better C, it offers better type checking and the opportunity to escape some of the danger points of its parent language e.g. defined constants and macros. In the context of numerical methods, the expectation is that C++ will allow a more highly structured approach than its parent language while retaining its efficiency. A matrix extension to C++ promises to give even greater functionality in carrying out matrix calculations. In particular it offers greater security than C and can incorporate the methods needed to overcome the technical weaknesses in C's handling of multi-dimensional arrays. Such extensions to a standard general purpose language offer greater openness and portability than characterize proprietary packages such as GAUSS.

C++ is a general purpose language that does *not* have an intrinsic matrix type, but its support for *classes* allows users to define their own types that have the same privileges as intrinsic types. With a C++ matrix class, the user has a system that combines the general power and flexibility of an increasingly popular general purpose language with the convenience offered by a matrix syntax. It is to be stressed that MatClass is not just for the experienced C++ programmer. Indeed, the combination of MatClass with friendly and sophisticated C++ development environments, such as that supplied by Borland, can be used as an introduction to programming. Those with some previous programming experience will be able to exploit the potential of a fully compiled language that offers a highly structured matrix syntax.

C++ uses *classes* to allow the user to extend the range of objects supported by the language and to facilitate structured programming methods and the use of *abstract data types*. Instances of classes are considered to be abstract *objects* whose internal structure need not be considered in order to successfully exploit their *behaviour*. For example, the typical user of MatClass need not be concerned with the detailed structure of a matrix object as long as it behaves properly e.g. they can be multiplied or added. From this perspective, it is usual to view a program as a sequence of instructions to objects to undertake various acts, e.g. print themsleves to a file. MatClass is largely made up of such *object methods*, so its user must get used to the syntax of object methods rather than the more traditional functional or procedural languages. Thus consider the use of matrix decompositions in MatClass. While one can solve equations with a "projection" operator, there are several classes of objects that encapsulate the services of matrix decompositions. To form the LU decomposition of a matrix, for example, one constructs a luDec object `L` from the matrix `A` with the statement `luDec L(A)`. Thereafter `L` is a LU decomposition of `A` which can be used to solve the equations `Ax=b` with the statement `L.solve(b)` or to estimate the condition number of `A` with `L.cond()`. Objects should be viewed, then, as active, not passive data structures. Indeed objects are a mixture of data and functions that offer the user various services.

While the ability to encapsulate an abstract data type, such as a matrix, is useful, our main interest in C++ is its support for inheritance. The user defined classes in C++ can *inherit* properties from previously defined classes allowing a *hierarchy* of object types. This is the key to object-oriented programming, to which we turn in the next section.

MatClass aims to lay the foundation for serious scientific work in C++ by being *open*, *extensible* and *portable*. It is *open* in the sense that the source code will be placed in the public domain. MatClass is *extensible* in the sense that the programmer can define his own classes of objects that inherit properties of the intrinsic classes and exploit the object and error management facilities of MatClass. Further user written extensions will be as efficient as MatClass intrinsics, this being in contrast to interpreted or semi-compiled systems such as GAUSS. MatClass is *portable* in that it is written solely in C++, involves no machine code or dependencies, and currently compiles correctly with PC compilers from Microsoft, Borland, Glockenspiel, Zortech and JPI, as well as Hewlett Packard's version of AT&T's *cfront* on their HP Series 700 Unix workstations. This portability gives the user the ability to move programs on to the system best suiting his needs. While MATLAB has been seen as an ideal system for *prototyping* algorithms before translating them to FORTRAN, the distinction between prototyping and development is blurred with MatClass. While users of GAUSS tend to see it as a self contained system isolating them from the dreaded compilers, it is not clear that sufficient consideration has been given to the deficiencies of GAUSS and to the attractiveness of C++ as a developer system. See section 9 for a discussion of the relative runtime efficiencies of GAUSS and MatClass.

One further aim of MatClass is to give the user ready access to state of the art numerical methods. The current structure of MatClass reflects an interest in implementating the key matrix decompositions and for solving equations and least-squares problems. Thus, the two main families of classes, matDec and matOls, are effectively "packages" built around the LU, Cholesky, Householder QR, and singular value decompositions. Key references are Press *et al*'s *Numerical Recipes in C* [13] and Golub and Loan's *Matrix Computations* [7]. Under the influence of the numerical methods literature some emphasis is placed on condition numbers, and both families give the user some form of estimating the condition number. In the same spirit, MatClass aims to give full support to singular-value (SVD) methods, particularly for least squares problems. Singular values not only have a ready interpretation as measures of "near singularity", but the SVD algorithms are numerically superior when the problem is effectively rank deficient. SVD methods have not being given the role in the econometrics literature they deserve, particularly when discussing rank and multicollinearity. The SVD should be readily available to all econometricians and even if it is felt that the computational costs of SVD are too high for routine work, it should nevertheless be possible to estimate the conditioning of the problem reliably, regardless of the underlying algorithm. If ill-conditioning is suggested then the worker can switch to SVD methods — or reconsider the structure of the model! It has to be noted that from this perspective GAUSS scores highly when compared with standard statistical packages.

In summary MatClass aims to provide services similar to those of GAUSS and MATLAB by giving the user a matrix syntax with access to state of the art numerical methods. What distinguishes MatClass from these others is that these services are provided in the framework of a compiled object-oriented environment and the full source code is to be placed in the public domain, a combination that allows third parties to correct and improve the code without runtime penalties. The object-oriented nature of the underlying compiler also allows a hierarchy of services to be developed in an efficient and systematic manner. To add new features to the system, the user does not have to rewrite the existing "package", his new offerings simply *inherit* and build on the features of the existing system. It is this aspect of MatClass to which we now turn.

4. OBJECT-ORIENTED NUMERICAL METHODS

Object-Oriented Programming Systems (OOPS) in general, and C++ in particular, have received a good deal of attention in the computing world. Two important features of OOPS are *encapsulation* and *inheritance*. Encapsulation refers to the ability to hide much of the detail of a data structure or an operation. While this notion has long been behind the structured approach to programming, its usage is greatly facilitated by OOPS. Inheritance refers to the ability of a new object type to inherit the properties of an existing type. When adding a new, or replacing an old, feature of an existing class, one simply extends or modifies rather than rebuilds the existing classes. In this way a layered structure of classes can be built on top of the fundamental management classes and the basic numerical procedures, allowing users to choose their own level of access. These features are important in the development of complex and co-operative software, particularly when the development involves more than one programmer. It will be argued that C++ with appropriate classes promises to be an efficient, highly structured environment in which to develop facilities and packages either individually or jointly.

The essential element of an object-oriented programming systems is its support for classes of objects that can encapsulate data types and inherit properties. To understand these concepts better and appreciate the arguments favouring their use, the reader is refered to Grady Booch's *Object-Oriented Design with Applications*[1]. Booch argues that the use of OOPS offers significant improvements over the more familiar structured programming approaches, particularly for complex "industrial-strength" software. Booch's discussion is directed to software houses which have to manage a team of programmers on a large project, but it is equally valid for a community of professionals developing a common system.

Booch (p.77) offers the following definition of an object:

> *An object has state, behaviour, and identity; the structure and behaviour of similar objects are defined in their common class; the terms instance and object are interchangeable.*

An object state will include not only the *external* state as perceived by the user of the object but also an *internal* state that reflects the details of the class implementation. For example, the external state of a matrix includes its dimensions (number of rows and columns) and the contents of its elements. In MatClass the internal state includes a 'map' which is used to translate references to matrix elements into memory addresses.

As part of the behaviour of a matrix we would certainly include the ability to be used in arithmetic expressions, e.g. be added to or multiplied by other, conformable matrices. From from the perspective of the user of a class, it is important that objects in the class behave in a valid and readily understood manner. The user should not typically have to consider the details of the implementation in order to understand the usage of the class — although occasionally efficiency may demand considered usage of a class. Indeed, it is highly desirable that the details be hidden from the users, restricting their access to a well defined interface. Such data and method hiding, i.e. encapsulation, allows the developer of the class to modify the implementation without upsetting other modules or user programs. For example, MatClass allocates storage for matrices column by column. It is relatively straightforward to modify a few key procedures to allocate the whole matrix in a single contiguous block of memory without changing the way in which matrices are used. Users of the class would continue to address individual elements of the class in the same manner, i.e. `A(i,j)` would still refer to the i-j*th* element of the matrix `A`.

A more substantial example would be a change in the algorithm used to form the LU decomposition of a matrix in the luDec class. MatClass comes with a luDec class that encapsulates the idea of an LU decomposition. Essentially, a luDec's internal state includes a matrix, a pivotal map and a status value. Normally a luDec object is used without explicit consideration of its state; rather the interest is in its behaviour — in the way it makes it easier to exploit the use of the decomposition to solve equations or estimate condition numbers. Having assigned a matrix to a luDec object, the user can request the object to estimate the condition number, test for singularity, and solve equations, assured that the object is handling the various housekeeping tasks and is *not* repeating significant calculations. In short, a luDec object encapsulates the *numerical procedures* that surround an LU decomposition. To declare and use a luDec object is to choose a set of numerical procedures with the common element of an LU decomposition. Currently, the luDec class uses a version of the Crout algorithm with partial pivoting to form the decomposition. It would not be difficult to replace this procedure with another variant of Gaussian Elimination without changing the external behaviour or usage of the class.

OOPS goes beyond encapsulation with the concept of **inheritance**, and it is worth trying to illustrate how this can be exploited when building numerical objects. MatClass has an abstract class of matrix decompositions, matDec. This class is abstract in that it does not actually encapsulate any decomposition method; it has purely abstract decomposition and solution methods. The four concrete decomposition classes — luDec, cholDec, qrhDec and svdDec — are *derived* from matDec, and all have actual decomposition and solution methods. As classes derived from matDec, these methods replace the abstract methods of matDec. More interestingly they inherit from matDec the ability to estimate condition numbers, solve multiple sets of equations, and form matrix inverses — not to mention various internal states and management functions. For example, a matDec object uses Hager's algorithm to estimate condition numbers. This method assumes the object to which it is applied knows how to solve equations. Abstract matDecs cannot solve equations, but concrete matDecs in the form of luDecs, cholDecs or qrhDecs can solve equations, and instances of these three classes use the same matDec method to estimate the condition number. It has to be stressed that the three concrete classes luDec, qrhDec and cholDec use the same piece of code to estimate condition numbers even though they have different methods for solving equations, and *this common piece of code does* not *have to be aware of all the possible classes that it will be servicing.* The condition-number method is being applied to an object, and each object knows what solution method to use on itself. Thus when the condition-number method needs to solve an equation, it effectively requests the object to apply its solution method to that equation. In C++ parlance, the solving of equations is a *virtual method*; the actual method is determined dynamically at runtime by the specific object to which the method is applied. In this way the ability to calculate condition numbers is inherited from the abstract matDec class by the three classes luDecs, cholDec and qrhDec. And further, this condition-number method can be inherited, without modification, by future derived classes that can solve equations, or if need be, it can be overridden as it is in the svdDec class.

As a second example, consider *refMatrix*, a concept that embodies the idea of a reference matrix — an object that is used to reference the whole or part of some other matrix. Thus, given a matrix `A` we can declare a refMatrix `col` to be used to refer to columns of `A`. Having attached `col` to `A`, we can instruct `col` to reference a specific column of `A`, say the *ith*, with `col.refCol(i)`. Thereafter `col` will act *as if* it were the *ith* column of `A`— at least until it is instructed to refer to some other part of `A`. For example the statement `col = 1` would assign the value 1 to the elements of the *ith* column of `A`. A reference matrix differs from a standard matrix in having to contain a reference to its underlying matrix, for example `col` holds an internal reference to `A`, and so it can be instructed to refer to various parts of that underlying matrix, for example `col` could be used to step through the columns of `A`. Otherwise a refMatrix *inherits all the properties of a standard matrix*. The refMatrix class is derived from the matrix class and any instance of refMatrix *is* a matrix and can be used wherever a matrix can be used.

The reader can no doubt imagine objects that are not matrices but which, on occasion, should behave as matrices. In statistical applications, for example, it can be useful to have the concept of a *data set* which can act like a matrix (of variables or cases), but can also handle variable and/or case names and other sample information such as missing values and periodicity. Likewise, a *table* is a useful output device that may take the form of a matrix of numbers with column and row labels — think of the standard table of estimated coefficients, standard errors and t-values. Here we would wish to manipulate the numerical fields like a matrix while having additional features for handing labels and for displaying of the labels and matrix.

It has to be stressed that the process of making one class inherit the properties of another is simple and relatively automatic. Virtual methods allow this process to go a stage further. With virtual methods a child can chose to override those methods of the parent that are unsuitable for its needs, and the child need not inherit all the features of the parent. For example, MatClass matrices can be reset in the sense that their dimensions can be modified dynamically. This would not be appropriate for a refMatrix, so the reset method is made virtual so that the refMatrix version can treat a call to reset as a fatal error. Similarly the concrete svdDec class overrides its MatDec parent's condition number method.

5. THE MATCLASS FAMILY

The foregoing discussion has shown that the key to OOPs programming is the implementation and use of appropriate classes of objects. Table 1 lists the classes currently offered by MatClass.

MatClass aims both to smooth the path for new programmers in writing productive code using C++ and to offer support for longer term development. To assist the newcomer, MatClass offers a straightforward syntax that does not require an understanding of the power and subtlety of C++. Thus MatClass assists with the management of objects and errors, offers a simple input-output system as well as a rich set of matrix intrinsics, supports the popular matrix decompositions, and offers a family of OLS classes on which future extensions can be built. At the same

TABLE 1
Main MatClass classes.

Class Name	Role
inFile	Input files
outFile	Output files
matObject	Container class used for errors and lists
matFunc	Function class used for errors and tracing
charArray	Arrays of characters
indexArray	Arrays of INDEXs
realArray	Simple arrays of REALs
matMap	Mappings of matrices into realArrays
matPair	Iterator for matrices
matrix	Central matrix class
refMatrix	Matrices that refer to some other matrix
matDec	Abstract class of matrix decompositions
luDec	LU decomposition class
cholDec	Cholesky decomposition class
qrhDec	Householder QR decompositions
svdDec	Singular Value Decompositions
matOls	Abstract OLS class
olsChol	OLS class based on Cholesky decomposition
olsQrh	QR based OLS class using Householder reflections
olsSvd	SVD based OLS class
matRandom	Random number generators
matSpecFunc	Abstract Special Function
logGammaFunc *etc*	Concrete Special Functions

time, the combination of a rich matrix class with an extensible object and error-management system gives the more serious user a powerful development tool for numerical research. It is likely that experienced programmers who have not used object-oriented systems will suffer from some conversion pains — this author certainly did! Object-oriented programming styles are very different from traditional procedural programming styles. Nevertheless the benefits are significant.

Alongside the central matrix class there is a number of general purpose classes as well as higher level families of classes related to the concept of a matrix. One

TABLE 2
MatClass scalar types.

Type Name	Role
INDEX	Unsigned integer, indices
REAL	Standard floating point
DOUBLE	Double precision floating point type
matError	Enumerated type for common errors

such family is based on various *matrix decompositions*, including the LU, Cholesky, QR by Householder reflections and SVD decompositions. Furthermore, there is a family of *ordinary least squares* (OLS) classes based on the Cholesky, QRH and SVD decompositions. Finally there are classes to support random number generation and the use of special matrix functions.

To support these central classes MatClass provides several underlying and subsidiary classes of objects. Apart from a concept of a matrix, which closely emulates the mathematical idea, MatClass gives support to scalar variables, including real numbers and integers, character arrays, and arrays of indices. MatClass uses the terms `REAL` and `DOUBLE` to define real variables. The `INDEX` type is an unsigned integer and is used primarily for index variables that control the addressing of matrix elements. The `LONG` is an unsigned long integer that is used where an `INDEX` variable will overflow.

MatClass supports *string constants* and introduces a class of `charArrays`. A sequence of characters placed between double quotes such as `"this is a string"` constitutes a string constant. Variables of type `charArray` can be declared and used to store strings of characters. For example, the statement `charArray name(40)` declares a variable `name` that can store a maximum of 40 characters, although the contents of a charArray variable may be less than its maximum length. These `charArrays` are particularly useful when reading strings from files.

The classes realArray and matMap are largely for internal use and offer the end user few services.

The use of classes to implement random number generators and special functions may seem surprising to those familiar with traditional programming methods. But objects need not simply be data structures; rather they are structures which, through their methods, offer a set of services to the user. A matRandom object not only has a state (the current state of the random generator) but also offers a number of methods to fill a matrix with random numbers and having a class of these objects affords the user access to several independent generators. Likewise, most of the special functions are implemented as classes, for example, there is a class logGammaFunc that is the basis of the logGamma function. These special function classes are derived from the abstract class matSpecFunc that offers its children a number of services. Thus all special *matrix* functions are driven by a procedure in the parent class. While the logGammaFunc class provides the specific code to return the value of logGamma

```
#include "matmath.hpp"
#include "olssvd.hpp"
#include "olschol.hpp"

void results( matOls& ols, matrix& y, matrix& x )
{
   const INDEX width = 10 ;
   matrix beta, stderr, tvalues, resid, fitted, newX ;

   ols.assign( y, x ) ;
   ols.coeff( beta ) ;
   ols.stdErr( stderr ) ;
   tvalues = beta.divij( stderr ) ;

   out.newLine() ;
   out( "Coeffs", width )( "Std Errors", width+2 ) ;
   out( "t-values", width+2 ).newLine() ;
   out( "----------", width )( "----------", width+2 ) ;
   out( "----------", width+2 ) ;
   matFormat( STACKED ) ;
   matField( width ) ;
   ( beta | stderr | tvalues ).put() ;
   out.newLine() ;

   out( "RSS    : " )( ols.rss() ).newLine() ;
   out( "TSS    : " )( ols.tss() ).newLine() ;
   out( "SE     : " )( ols.se() ).newLine() ;
   out( "RSQ    : " )( ols.rsq() ).newLine() ;
   out( "RBarSq : " )( ols.rBarSq() ).newLine() ;
   out( "DW     : " )( ols.dw() ).newLine() ;
   out( "Cond   : " )( ols.cond() ).newLine() ;

} // results
```

Fig. 1. Using OLS Classes Part A.

for some real argument, the matrix version of logGamma is essentially governed by code in the parent class matSpecFunc. In this way special real-valued functions are converted into matrix functions without undue duplication of requisite loops and error checking.

6. AN OLS EXAMPLE

Figures 1 and 2 list the two parts of a program illustrating the use of two OLS classes, olsSvd and olsChol, that are based on SVD and Cholesky decompositions,

```
main()
{
   INDEX M, N ;
   inFile data ;

   out( "\n\nTest of ols\n\n" ) ;

   data.open( "jsex3n.dat" ) ;
   data(M)(N) ;

   matrix X(M,N), c(M-1), x, Y(M), y, sv ;

   X.get( data ) ;
   Y.get( data ) ;

   X = ln( X ) ;
   Y = ln( Y ) ;

   c = 1.0 ;
   y = Y.smpl( 2, M ) ;
   x = c | X.smpl( 2, M ) | Y.smpl( 1, M-1 ) ;

   olsSvd ols1 ;
   out( "\nResults from olsSvd : \n" ) ;
   results( ols1, y, x ) ;

   olsChol ols2 ;
   out( "\nResults from olsChol : \n" ) ;
   results( ols2, y, x ) ;

   return 0 ;
} // main
```

Fig. 2. Using OLS Classes Part B.

respectively. The program is made up of two functions, `results` and `main`. `main` is always the entry point into a C++ program, so that execution starts with the lines declaring the two INDEX variables `M` and `N`. The main function then reads in a data set, applies log transforms, and builds a regressand and a matrix of regressors including a lagged regressand. Having created an olsSvd object `ols1`, it calls `results` to print out some of the standard OLS results. This is repeated for an olsChol object `ols2`. The output from the program is given in Figure 3.

This example illustrates some of the features of MatClass, but more interestingly it also illustrates one of the uses of class hierarchies. Looking closely at the `results`

```
Results from olsSvd :

    Coeffs  Std Errors    t-values
----------  ----------  ----------
   0.06734     0.01951       3.451
   -0.0145     0.01995     -0.7269
  -0.02009     0.01986      -1.011
     1.028     0.01281       80.22

 RSS    :    0.05518
 TSS    :  2.414e+04
 SE     :    0.04795
 RSQ    :          1
 RBarSq :          1
 DW     :     0.7388
 Cond   :      7.355

Results from olsChol :

    Coeffs  Std Errors    t-values
----------  ----------  ----------
   0.06734     0.01951       3.451
   -0.0145     0.01995     -0.7269
  -0.02009     0.01986      -1.011
     1.028     0.01281       80.22

 RSS    :    0.05518
 TSS    :  2.414e+04
 SE     :    0.04795
 RSQ    :          1
 RBarSq :          1
 DW     :     0.7388
 Cond   :      10.66
```

Fig. 3. OLS Output.

function in Figure 1, you will see that its first argument is declared to be a matOls, not an olsSvd or olsChol. Despite C++'s being a strongly typed language, the compiler does *not* treat the calls to `results` with objects of type olsSvd and olsChol as errors, since these objects *are* matOls objects. Instances of derived classes are instances of the parent class. We have a family of matOls classes, and any object in those classes can be used where a matOls object is expected. Thus we only need to write *one* `results` function for all existing and future matOls classes. Thinking in terms of *families* of classes allows us to introduce a higher level of abstraction into our programs, giving greater freedom for mixing and matching objects to specific needs. Although not part of our example, the user can choose the specific class, and thus algorithm, used by `results` at runtime.

This example hints at what is possible when building a hierarchy of classes and illustrates the fact that many of the benefits of inheritance can be exploited by stand-alone functions such as `results`. Although MatClass does not currently offer classes for nonlinear models, these featues of C++ are expected to be particularly valuable in this area.

7. ERRORS IN MATCLASS

MatClass maintains a stack of function calls and displays a list of the active functions when an error occurs. For a function to appear on this stack, it must declare a variable of type matFunc. All major functions in MatClass exploit this facility and thus it should be possible to identify the sequence of calls that led to the error.

Furthermore an error message will normally induce MatClass to generate a list of objects. MatClass objects arrange themselves in "levels" for lists, the matrix class itself is at level 3. The MatClass object lists can be restricted to "high" level objects through the `matListCtrl` function.

Unfortunately, MatClass has no access to the C++ user defined identifiers, unless the program explicitly names objects (using the `name` method). We eschew the details of the object lists here, simply noting that each object attempts to provide summary information on its state.

8. TRACING PROGRAMS WITH MATCLASS

While most modern C++ compilers come with source level debuggers, and some come with class browsers, I have found it useful to have MatClass-specific, debugging facilities. This can be controlled by setting the depth of the trace. The default level is zero which means no tracing takes place.

During a trace, each major MatClass function occuring at or below the trace depth identifies itself by name. This is normally followed by information on the principal object with which the function is working. Tracing can be switched on anywhere in a program by setting a nonzero debug level and can be switched off by setting a zero debug level. By default the debug output is written to the standard output file but can be redirected to a disk file.

9. EFFICIENCY OF MATCLASS IN ACTION

This section compares GAUSS with the combination of MatClass and a C++ compiler as *development systems*. Currently GAUSS has advantages for those whose needs are met by the various "add-ons" such as GAUSSX. MatClass cannot yet satisfy these needs, but there is a commitment toward providing "higher" level capabilities in future versions.

- While the structure of GAUSS has advantages for "interactive" work and can form the basis for significant research, it is not as well structured as C++. In particular, it suffers from not being strongly typed and not offering object-oriented facilities. As we have illustrated the extensibility characteristic of MatClass makes it superior to traditional languages such as GAUSS.
- In assessing MatClass, one is also assessing in part the available compilers. The current version of MatClass compiles and runs with all compilers available to the author: Glockenspiel version v2.0 d2, Borland's C++ version 3.1, Microsoft C/C++ v7, Zortech v2.1, JPI v1, and Hewlatt Packard's CC compiler provided with their series 700 workstations. The Glockenspiel and HP compilers use the AT&T Unix-based *cfront*, and so it is expected that the class will readily be ported to this and related systems.

  Compiler systems are varied and their appeal will differ according to the background of the user. Experienced programmers will appreciate using Glockenspiel and other command-orientated systems with familiar development tools. We shall see that the Glockenspiel system produces very efficient code — in fact the machine code is generated by Microsoft's C compiler. By contrast, 'integrated" environments, such as those offered by Borland, will satisfy those who want the compilation process to be straightforward and relatively automatic. Anyone who is using systems such as GAUSS to produce programs of any significant size will find the move to this environment relatively painless and productive. Not only is the compilation process easy, but you also have a first class editor and debugger.[4]
- The fact that MatClass works with these compilers suggests that any MatClass code can be readily ported across operating systems. Glockenspiel, Zortech and JPI have versions of their compilers for OS/2 — a much maligned system that *is* better than DOS — and Borland is flagged by IBM to be a major player in the future. All DOS based compilers offer support for MS-Windows and several come with optional DOS extenders.
- The reader should not view MatClass as an attempt to give the user a total and final solution for his computational work; rather it is an attempt to provide one development path based on the potential of C++. There are already alternatives, for example M++ from Dyad Software, and I am confident there will be others.
- One price to be paid for using a C++ compiler rather than GAUSS's pseudo-compiler[5] is the relative slow compiling. GAUSS has rightly gained a reputation

[4] You also get command line versions of the compiler and make facility allowing you to base your work on editors such as Brief or SPE.

[5] GAUSS produces some form of pseudo-code rather than linkable machine object code.

TABLE 3
Timings in seconds for 90 × 90 multiplication

System	Machine	Compile	Initial	Multiply	Code Size K
GAUSS v2	486 PC	NA	3.02	0.77	NA
GAUSS386	486 PC	NA	2.34	0.62	NA
Glockenspiel v2	486 PC	12.9	0.05	1.87	85
Microsoft v7	486 PC	7.6	0.06	2.03	80
Borland v3.1	486 PC	2.8	0.05	2.08	84
HP CC	HP 720	2.3	0.01	0.11	106

for its speed of execution, and for some tasks this can be a critical factor. But it has to be understood that this feature is only true of its intrinsic operations such as matrix multiplication. When using low level code such as addressing individual elements of matrices, however, the interpretitive nature of GAUSS becomes highly inefficient. In any event the combination of MatClass with a compiler system such as Glockenspiel competes well with GAUSS. Some evidence on these issues is given below.

As a crude measure of relative efficiency Table 3 reports timings for the simple task of multiplying two 90 by 90 matrices. Apart from forming the actual product, this job involves initialising the two matrices with a nested pair of loops that step through the rows and columns. This initialisation allows comparison of the relative efficiencies in low-level access to the elements of the matrices.

- Two PC configurations were used, an Opus and a DAN. Both used a 80486 at 33Mhz with 8Mbs of RAM and a 64K cache. The DAN had a 2Mb hard disk cache. The timings for the GAUSS runs were based on the Opus, while those for the C++ compilers were based on the DAN. The presence of the hard disk cache significantly improved the compilation times. Both machines used DOS 5. It is fair to say that a minimum requirement for making good use of C++, and implicitly MatClass, under DOS is a machine based on the 80386DX or 80386SX with a reasonable hard disk.
- The timings for the HP CC runs were based on a Hewlett-Packard 720 workstation running a PA-RISC processor at 50Mhz, with 128k of instructor cache, 256k of data cache, 32 megabytes of RAM, and running HP-UX. The compilation and execution of the programs was done from the shell of GNU Emacs. Without placing too much weight on one set of timings, these so-called "Snakes" are truly impressive and bear witness to the possibilities offered by RISC. Some readers may question the value of raw speed without general productivity tools or may not see Unix workstations as their preferred platform, but I have found the system with CC, GNU Emacs, TEX and MatClass to be highly productive. As the prices of Unix workstations and higher level PCs continue to fall, machines of this calibre will soon be the standard for research. With such configurations

significance of compiler and linker lags largely disappear. Note the total times for the HP is less than the total for GAUSS386. For more computationally tasks the sheer processing power of the HP is even more important. All in all, it is suggested that the ability to port code from PCs to powerful RISC systems is a significant advantage of MatClass.

- All compilers were set for speed optimisation — in the case of Glockenspiel and Microsoft the options –Ox –Op were used. Glockenspiel translates C++ into C and then calls the Microsoft C v7 compiler to generate machine code. The full optimisation option is the -Ox -Op option for the C compiler. In the case of Borland, the command line compiler was used with the options –O2.
- The reported compilation times are averages of 5 runs clocked with a stop watch. No attempt was made to time GAUSS' compilation, partly because there is no clear indication when the compilation is complete and partly because it seems to be negligible for this small job. The timings for initialisation and multiplication were generated directly by the code using calls to the system clock.
- The execution times for the multiplication suggest that the advantages of GAUSS' assembler code are real but not overwhelming when compared with modern day optimising compilers. Although the performance of the GAUSS intrinsics on the 486 is particularly impressive, the times for the Glockenspiel/Microsoft combination are equally so for code written in an object-oriented system. Given the advantages of developing code in a higher level language the argument for dropping down to assembler is not strong. It has to be stressed that the implementation of matrix multiplication in MatClass does *not* involve any fancy tricks.
- Comparing Glockenspiel and GAUSS confirms the trade-off between compilation times and execution speeds for low-level code. Consideration of the initialisation times suggests the running of low-level code in GAUSS is very slow when compared to fully compiled code. The conclusion is that GAUSS gives good execution times as long as the main computations can be completed using its intrinsics, but if one needs any low-level code, then GAUSS will not give efficient run times. In particular, it is not feasible to substitute your own GAUSS routines for GAUSS intrinsic functions without severe loss of efficiency. In MatClass there is no runtime penalty for replacing or adding new low-level functions.
- The Glockenspiel V1.2 and Microsoft C V5.1 combination previously set the standard for quality compilation of C++ on PCs. The consistency of this combination has been impressive and the computational performance first class. While version 2 of Glockenspiel still generates fine code and has the advantage of being a port of *cfront*, the native compilers such as Microsoft v7 and Borland v3.1 have developed into fine products.
- Turning to the performance of the Borland compiler, it can be seen that the timings are almost as good as Microsoft. In the past the floating-point performance of the Borland compilers has not been impressive. With this compiler they have caught up. Furthermore their development environment is easier to use than Microsoft's workbench – although I have been able to configure the latter better

for my needs.

- A comment may be offered on the sizes of compiled code. There is clearly a danger that significant applications based on MatClass will soon run into memory difficulties on PCs running DOS. There is a number of ways of overcoming this limitation. Most of the compilers have a version that supports, even if they do not provide, a DOS extender and this is likely to be best solution. And there are smart linkers which do not link redundant code; for example the JPI DOS compiler's version of the multiply program was approximately 48K, half the size of the others. One interesting possibility is promised by the new Microsoft C/C++ version 7 compiler that allows the use of *p*-code for modules that are non time-critical. And of course there are overlay systems like Borland's "VROOM"! The size of the HP code is not vastly greater than the PC code despite it being based on RISC.

10. SYSTEM REQUIREMENTS

As a C++ class, the essential ingredient for the use of MatClass is a computer running a C++ compiler. The rapid adoption of the language by the industry means C++ is well supported on most popular systems, including PCs, Macintoshes, and Unix workstations. It is likely that the minimal requirement for any significant use will be a 386SX PC with a hard disk, or machines of comparable performance. The author must confess to doing some of the original development of MatClass using Glockenspiel on a humble domestic PC/XT clone, where compilation times were something of an impediment and nearly impossible with later compilers. With the current standard of a 486 machine with disk cache and graphics accelerator, however, these compiler lags are rapidly becoming insignificant.

11. AVAILABILITY

The source code of MatClass is available from the author. The source itself is in the public domain in the spirit of the Free Software Foundation; that is, the source code for MatClass will be covered by a free licence which safeguards the free availability of that code. My own experience with free software — Kermit, Emacs and TeX— suggests that it works best when someone publishes, in a traditional book form, a guide and manual. It is my intention to do this for MatClass and a draft of such a text is available. This is also available in electronic form — in raw TeX, DVI, HP or PostScript — but I retain the copyright to allow future publication. All of the MatClass files are available from the UTS machine at the Manchester Computing Centre using an anonymous ftp to `uts.mcc.ac.uk`. Look in the `pub` subdirectory for `matclass`. Alternatively the files can be supplied on disk direct form the author at a cost to cover expenses.

As free software, there will be no commercial warranty. Nor can I offer free support as a right. Clearly I wish to make MatClass useful and as correct as possible, and toward this end I intend to respond to problems and bug reports as well as I can within the available resources. I will try to support the use of MatClass with Microsoft C/C++ v7 and Borland C++ Version 3.1 for DOS and HP's CC on their series 700 workstations. I will also be able to give limited support to Glockenspiel C++ Version 2.0d2 for DOS, but beyond that I can only make available pointers on loading MatClass on other systems.

12. BIBLIOGRAPHY

The full use of MatClass requires some understanding of C++. Probably the best overall introduction is Stanley B. Lippman's *C++ Primer* [10]. Beyond that, consider *Programming in C++* by Stephen C. Dewhurst and Kathy T. Stark [3]. Look to Lippman for further information on control structures such as `for` loops and `if` statements and for the general structure of functions and methods in C++. Dewhurst and Stark give an excellent introduction to the potential offered by C++'s object orientation. To go further yet see the excellent *Advanced C++ Programming Styles and Idioms* by Coplien [2].

The two "bibles" on C++ are authored by the originator of the langauge, Bjarne Stroustrup. *The C++ programming Language*[14] was the effective definition of the first version of the language. With coauthor Margaret Ellis, *The Annotated C++ Reference Manual* [6] is a draft of an ANSI standard definition of the second version of the language. Neither of these texts is for the casual reader.

The author must acknowledge the influence of Bruce Eckel's *Using C++* [5] and Scott Ladd's *C++ : Techniques and Applications* [9].

See Booch [1] and Meyer [11] for a discussion of object-oriented software design.

REFERENCES

1. G. Booch. *Object-Oriented Design with Applications*, Benjamin/Cummings, Redwood, 1991.
2. J.O. Coplien. *Advanced C++ Programming Styles and Idioms*, Addison-Wesley, Reading Mass., 1992.
3. S.C. Dewhurst and K.T. Stark. *Programming in C++*, Prentice Hall, Englewood Cliffs, 1989.
4. J.J. Dongarra, C.B. Moler, J.R. Bunch and G.W. Stewart. *LINPACK User's Guide*, SIAM, Philadelphia, 1979.
5. B. Eckel. *Using C++*, Osborne McGraw-Hill, Berkeley, 1989.
6. M.A. Ellis and B. Stroustrup. *The Annotated C++ Reference Manual*, Addison-Wesley, Reading Mass., 1990.
7. G.H. Golub and C.F. van Loan. *Matrix Computations*, John Hopkins, Baltimore, 1989.
8. P. Griffiths and I.D. Hill (editors). *Applied Statistics Algorithms*, Ellis Horwood, 1985.
9. S. Ladd. *C++ Techniques and Applications*, Prentice-Hall, New York, 1990.
10. S.B. Lippman. *C++ Primer*, First Edition, Addison-Wesley, Reading Mass., 1989.
11. B. Meyer. *Object-oriented Software Construction*, Prentice Hall, UK, 1988.
12. W.H. Press, B.P. Flannery, S.A. Teukolsky and W.T. Vetterling. *Numerical Recipes*, Cambridge University Press, Cambridge, 1986.

13. W.H. Press, B.P. Flannery, S.A. Teukolsky and W.T. Vetterling. *Numerical Recipes in C*, Cambridge University Press, Cambridge, 1988.
14. B. Stroustrup. *The C++ Programming Language*, First Edition, Addison-Wesley, Reading Mass., 1986.

ISMAIL CHABINI, OMAR DRISSI-KAÏTOUNI AND MICHAEL FLORIAN

Parallel Implementations of Primal and Dual Algorithms For Matrix Balancing

ABSTRACT. We report the parallel computing implementations of a primal projected gradient algorithm and the classical RAS dual algorithm for matrix balancing. The computing platform used is a network of Transputers which is suitable for coarse grained parallelization of sequential algorithms. We report computational results with dense matrices of dimension up to 315×315 and $100,000$ nonzero variables.

1. INTRODUCTION

The recent development and use of computing platforms based on parallel processing architecture has had a major impact on many fields of science and economics that require intensive computations (Bertsekas and Tsitsiklis, 1989; Zenios, 1989; Pardalos et al., 1990). The efficient use of various parallel computing architectures requires the development of new code that takes advantage of the development tools and compilers that are available. In this paper, we report on parallel implementations of primal and dual algorithms for matrix balancing problems on a network of Transputers. This is probably one of the least costly Multiple Instruction Multiple Data (MIMD) (Flynn, 1972) parallel computing platforms available and is best suited for "coarse grain" parallelization of sequential codes. Nevertheless, our experiments indicate that significant gains in speed and efficiency accompany its use when compared to the execution of the sequential code on a single Transputer.

This paper is organized as follows. In the following Section, the matrix balancing problem is described. Sections 3 and 4 present primal and dual algorithms for this problem. Then, the parallel versions of these algorithms are described in Section 5. The computational results obtained and their evaluation are given in Section 6. Our conclusions and views on further development of parallel computing implementations of matrix balancing problems comprise the last Section.

2. THE MATRIX BALANCING PROBLEM

The matrix balancing problem that we consider can be defined as follows: given an $n \times m$ nonnegative matrix g^0, supply and demand vectors O and D of dimensions n and m, respectively, an $n \times m$ matrix g^* is sought that satisfies

D. A. Belsley (ed.), Computational Techniques for Econometrics and Economic Analysis, 173–185.

$$\sum_{j=1}^{m} g_{ij}^{*} = O_i, \qquad i = 1, \ldots, n, \tag{1}$$

$$\sum_{i=1}^{n} g_{ij}^{*} = D_j, \qquad j = 1, \ldots, m, \tag{2}$$

$$g_{ij}^{*} = R_i g_{ij}^{0} S_j, \qquad i = 1, \ldots, n; \quad j = 1, \ldots, m, \tag{3}$$

$$\Big(\sum_{i=1}^{n} O_i = \sum_{j=1}^{m} D_j = T\Big).$$

This problem occurs frequently in economics, transportation planning, statistics, demographics and image reconstruction. A good survey of applications of matrix balancing may be found in Schneider and Zenios (1990).

Many algorithms have been constructed for matrix balancing problems. These may be viewed as primal or dual approaches for solving (1)–(3). The primal algorithm that we present in this paper is new and is based on an analytical gradient projection. It is well known that (1)–(3) are equivalent to the following entropy optimization problem:

$$\text{Min} \quad F(g) = \sum_{i=1}^{n} \sum_{j=1}^{m} g_{ij} \Big(\ln \frac{g_{ij}}{g_{ij}^{0}} - 1 \Big) \tag{4}$$

subject to (1) and (2).

The solution g_{ij}^{*} is nonzero; hence there is no need to state nonnegativity constraints on $[g]$.

3. THE PRIMAL ALGORITHM

The gradient projection method (Luenberger, 1984) is a primal nonlinear programming algorithm that has not been adapted so far for the problem (4), (1)–(2). Here we give its adaptation to the matrix balancing problem as well as some details of the proofs.

Consider $X = [x_{ij}]$ to be an $n \times m$ matrix satisfying the conservation of flow equations

$$\begin{aligned} -\sum_{j=1}^{m} x_{ij} &= 0, \qquad i = 1, \ldots, n, \\ \sum_{i=1}^{n} x_{ij} &= 0, \qquad j = 1, \ldots, m-1, \end{aligned} \tag{5}$$

where the constraint $\sum_{i=1}^{n} x_{im}$ has been removed, since (5) is a linear system of rank $(n + m - 1)$ and is redundant. By using the lexicographic ordering of the subscripts

i and j, we can convert the $n \times m$ matrix X into an $n \cdot m$ vector x. Let y be the orthogonal projection of x on the null space of the constraints (1), (2) defined by (5). y is the solution to the problem

$$\text{Min} \quad \tfrac{1}{2}\|y - x\|^2 \tag{6}$$

subject to (5), which may be restated as

$$Ay = 0, \tag{7}$$

where A is the node-arc incidence matrix with destination m removed.

The K-K-T necessary and sufficient conditions for (6)–(7) are

$$\begin{aligned} y - x + A^{\mathrm{T}}\lambda &= 0, \\ Ay &= 0. \end{aligned} \tag{8}$$

It is relatively easy to show that the vector of dual variables λ associated with the constraints of (7) is given by the formula

$$\lambda = (AA^{\mathrm{T}})^{-1}Ax. \tag{9}$$

Drissi (1991) was the first to find an explicit analytical expression for $(AA^{\mathrm{T}})^{-1}$ in order to compute λ. In the following, we develop another method for solving (9), which has the advantage of showing why (9) is easy to solve for bipartite networks. A byproduct of this analysis is an explicit expression of y as a function of x which is easy to manipulate. We present first some preliminary results.

Consider the following linear system:

$$\left[\begin{array}{c|c} A_{11} & A_{12} \\ \hline A_{21} & A_{22} \end{array}\right] \begin{bmatrix} X_1 \\ X_2 \end{bmatrix} = \begin{bmatrix} B_1 \\ B_2 \end{bmatrix}, \tag{10}$$

where X_1, X_2, B_1 and B_2 are column vectors, and A_{11}, A_{12}, A_{21} and A_{22} denote submatrices of appropriate dimensions.

The system (10) is equivalent to

$$QX_2 = R, \tag{11}$$

$$X_1 = A_{11}^{-1}(B_1 - A_{12}X_2), \tag{12}$$

where $Q = A_{22} - A_{21}A_{11}^{-1}A_{12}$ and $R = B_2 - A_{21}A_{11}^{-1}B_1$. If A_{11} is easily invertible, then the solution of (10) is equivalent to the solution of (11), which is a linear system of a lower dimension.

Consider now (AA^{T}) of (9). It may be expressed as

$$(AA^{\mathrm{T}}) = \left[\begin{array}{c|c} mI_{n,n} & -U_{n,m-1} \\ \hline -U_{m-1,n} & nI_{m-1,m-1} \end{array}\right], \tag{13}$$

where I is the identity matrix and U is the unitary matrix. Hence solving (9) is equivalent to finding a solution to the system

$$(AA^{\mathrm{T}})\begin{bmatrix}\lambda_1\\ \lambda_2\end{bmatrix} = \begin{bmatrix}[Ax]_1\\ [Ax]_2\end{bmatrix}.$$

The indices belonging to the set 1 correspond to origins and those belonging to set 2 to the $(m-1)$ destinations.

In the linear system (11), the matrix Q is given by

$$Q = (nI_{m-1,m-1} - (-U_{m-1,n})\left(\frac{1}{m} I_{n,n}\right)(-U_{n,m-1})),$$

$$Q = \left(nI_{m-1,m-1} - \frac{1}{m} U_{m-1,n}U_{n,m-1}\right),$$

$$Q = \left(nI_{m-1,m-1} - \frac{n}{m} U_{m-1,m-1}\right),$$

$$Q = n\left(I_{m-1,m-1} - \frac{1}{m} U_{m-1,m-1}\right),$$

$$Q = n\begin{bmatrix} 1-\frac{1}{m} & -\frac{1}{m} & \cdots & -\frac{1}{m}\\ -\frac{1}{m} & 1-\frac{1}{m} & \cdots & -\frac{1}{m}\\ \vdots & \vdots & \ddots & \vdots\\ -\frac{1}{m} & -\frac{1}{m} & \cdots & 1-\frac{1}{m}\end{bmatrix}.$$

We note that the columns of Q are permutations of the first one. This is due to the topology of the network. Since it is a complete bipartite network, the different destinations may be permuted without changing the "drawing" of the graph. Due to its special structure, Q has an easy inverse.

Proposition 1

$$Q^{-1} = \frac{1}{n}(I_{m-1,m-1} + U_{m-1,m-1}).$$

Proof:

$$QQ^{-1} = n\left(I_{m-1,m-1} - \frac{1}{m} U_{m-1,m-1}\right)\frac{1}{n}(I_{m-1,m-1} + U_{m-1,m-1}),$$

$$QQ^{-1} = \left(I_{m-1,m-1} - \frac{1}{m} U_{m-1,m-1} + U_{m-1,m-1} - \frac{1}{m} U_{m-1,m-1}U_{m-1,m-1}\right),$$

since $U_{m-1,m-1}U_{m-1,m-1} = (m-1)U_{m-1,m-1}$, it follows that

$$QQ^{-1} = \left(I_{m-1,m-1} + \frac{m-1}{m} U_{m-1,m-1} - \frac{1}{m}(m-1)U_{m-1,m-1}\right),$$

$$QQ^{-1} = I_{m-1,m-1}.$$

□

Proposition 2: The solution of (9) is given by the following expressions

$$\lambda_{1_i} = \frac{1}{nm} \sum_{k=1}^{m} \sum_{l=1}^{n} x_{lk} - \frac{1}{m} \sum_{k=1}^{m} x_{ik} - \frac{1}{n} \sum_{l=1}^{n} x_{lm} \quad i = 1, \ldots, n, \tag{14}$$

$$\lambda_{2_j} = \frac{1}{n} \left(\sum_{l=1}^{n} x_{lj} - \sum_{l=1}^{n} x_{lm} \right) \quad j = 1, \ldots, m. \tag{15}$$

Proof:
We evaluate R of (12) by using the above result.

$$R = [Ax]_2 - (-U_{m-1,n}) \times \left(\frac{1}{m} I_{n,n} \right) [Ax]_1,$$

$$R = [Ax]_2 + \frac{1}{m} U_{m-1,n} [Ax]_1.$$

Using (11) and the result of Proposition 1 it follows that

$$\lambda_2 = Q^{-1} R,$$

$$\lambda_2 = \frac{1}{n} (I_{m-1,m-1} + U_{m-1,m-1}) \left([Ax]_2 + \frac{1}{m} U_{m-1,n} [Ax]_1 \right),$$

$$\lambda_2 = \frac{1}{n} \left([Ax]_2 + U_{m-1,m-1} [Ax]_2 + \frac{1}{m} U_{m-1,n} [Ax]_1 + \frac{m-1}{m} U_{m-1,n} [Ax]_1 \right),$$

$$\lambda_2 = \frac{1}{n} ([Ax]_2 + U_{m-1,m-1} [Ax]_2 + U_{m-1,n} [Ax]_1).$$

For a destination node $j \in \{ 1, 2, \ldots, m-1 \}$

$$\lambda_{2_j} = \frac{1}{n} \left(\sum_{l=1}^{n} x_{lj} + \sum_{k=1}^{m-1} \sum_{l=1}^{n} x_{lk} + \sum_{l=1}^{n} \left(- \sum_{k=1}^{m} x_{lk} \right) \right),$$

$$\lambda_{2_j} = \frac{1}{n} \left(\sum_{l=1}^{n} x_{lj} + \sum_{k=1}^{m-1} \sum_{l=1}^{n} x_{lk} - \sum_{k=1}^{m-1} \sum_{l=1}^{n} x_{lk} - \sum_{l=1}^{n} x_{lm} \right),$$

$$\lambda_{2_j} = \frac{1}{n} \left(\sum_{l=1}^{n} x_{lj} - \sum_{l=1}^{n} x_{lm} \right).$$

We recall that the node $j = m$ corresponds to the redundant constraint that was dropped; its dual variable $\lambda_{2_m} = 0$. The expression (15) holds for $j = m$ as well. Hence

$$\lambda_{2_j} = \frac{1}{n} \left(\sum_{l=1}^{n} x_{lj} - \sum_{l=1}^{n} x_{lm} \right), \qquad j = 1, \ldots, m.$$

In order to evaluate λ_1, the vector of dual variables (prices) associated with source nodes, one can apply (12) to obtain

$$\lambda_1 = \frac{1}{m} I_{n,n}([Ax]_1 - (-U_{n,m-1})\lambda_2),$$

$$\lambda_1 = \frac{1}{m} ([Ax]_1 + U_{n,m-1}\lambda_2).$$

Let $i \in \{1, 2, \ldots, n\}$ represents an origin node

$$\lambda_{1_i} = \frac{1}{m} \Big(- \sum_{k=1}^{m} x_{ik} + \sum_{k=1}^{m-1} \lambda_{2_k} \Big).$$

Since $\lambda_{2_m} = 0$,

$$\lambda_{1_i} = \frac{1}{m} \Big(- \sum_{k=1}^{m} x_{ik} + \sum_{k=1}^{m} \lambda_{2_k} \Big). \tag{16}$$

By using (15),

$$\begin{aligned} \sum_{k=1}^{m} \lambda_{2_k} &= \sum_{k=1}^{m} \frac{1}{n} \Big(\sum_{l=1}^{n} x_{lk} - \sum_{l=1}^{n} x_{lm} \Big) \\ &= \frac{1}{n} \sum_{k=1}^{m} \sum_{l=1}^{n} x_{lk} - \frac{1}{n} \sum_{k=1}^{m} \sum_{l=1}^{n} x_{lm} \\ &= \frac{1}{n} \sum_{k=1}^{m} \sum_{l=1}^{n} x_{lk} - \frac{m}{n} \sum_{l=1}^{n} x_{lm}. \end{aligned}$$

By replacing $\sum_{k=1}^{m} \lambda_{2_k}$ with $\frac{1}{n} \sum_{k=1}^{m} \sum_{l=1}^{n} x_{lk} - \frac{m}{n} \sum_{l=1}^{n} x_{lm}$ in (16), one obtains (14). □

Now, we can obtain an explicit expression for y as a function of x. The linear system (8) implies that

$$y = x - A^{\mathrm{T}}\lambda,$$

$$y_{ij} = x_{ij} - (-\lambda_{1_i} + \lambda_{2_j}).$$

By replacing λ_{1_i} and λ_{2_j} by their expression (14) and (15), we obtain

$$y_{ij} = x_{ij} + \frac{\sum_{l=1}^{n} \sum_{k=1}^{m} x_{lk}}{nm} - \frac{\sum_{k=1}^{m} x_{ik}}{m} - \frac{\sum_{l=1}^{n} x_{lj}}{n}, \tag{17}$$

which proves the proposition below.

Proposition 3: The expression of y as a function of x is given by (17). □

The primal algorithm that is implied by these results may be stated as follows:

Projected Gradient Algorithm

Step 0 (Initialization):

$$g_{ij}^r = \frac{O_i D_j}{T}, \text{ for all } (i,j); \qquad r = 1$$

Step 1 (Computation of the gradient):

$$\nabla F_{ij}^r = \ln\left(\frac{g_{ij}^r}{g_{ij}^0}\right), \text{ for all } (i,j)$$

Step 2 (Computation of the projected gradient):

Apply equation (17) to obtain the projected gradient P^r.

Step 3 (Optimality test):

If $\|P^r\| < \epsilon$, STOP;
otherwise continue to step 4.

Step 4 (Line search):

$\alpha^r = \arg\min_{0 \leq \alpha \leq \alpha_{\max}} F(g^r - \alpha P^r)$
(The line search is performed with a Newton method)

Step 5 (Update variables):

$$g^{r+1} = g^r - \alpha^r P^r$$
$$r = r + 1 \quad \text{and return to step 1.}$$

4. THE DUAL METHOD FOR MATRIX BALANCING

The classical method for solving (1)–(3) is a dual method that dates back to 1937 (Kruithof, 1937) and was generalized for general linear constraints by Bregman (1967a, 1967b). It is used widely in transportation planning applications (Wilson, 1967; Evans, 1967; Robillard and Stewart, 1974). It is also known as the RAS algorithm (Bachem and Korte, 1979), since it alternates between scaling rows and columns, which is equivalent to premultiplying by R and postmultiplying by S to obtain the balanced matrix (R and S are diagonal matrices). The algorithm may be viewed as a coordinate ascent method for the Lagrangean dual of the entropy optimization problem [Cottle, Duval and Zihan (1986), Schneider and Zenios (1989)]. The algorithm may be stated as follows:

RAS Algorithm

Step 0 (Initialization):

$R^0 = 1; \qquad r = 1$

Step 1 (Balancing columns):

For each $j = 1, 2, \ldots, m$ DO:

$$S_j^r = \frac{D_j}{\sum_{i=1}^{n} R_i^{r-1} g_{ij}^0}$$

Step 2 (Balancing rows):

For each $i = 1, 2, \ldots, n$ DO:

$$R_i^r = \frac{O_i}{\sum_{j=1}^{m} S_j^r g_{ij}^0}$$

Step 3 (Stopping test):

If $\max_{1 \le i \le n} \left| \frac{R_i^r - R_i^{r-1}}{R_i^r} \right| \le \epsilon$ and $\max_{1 \le j \le m} \left| \frac{S_j^r - S_j^{r-1}}{S_j^r} \right| \le \epsilon$, go to step 4;
otherwise return to step 1.

Step 4 (Compute solution):

$g_{ij}^r = R_i^r g_{ij}^0 S_j^r, \qquad \forall (i, j)$ and STOP.

We stated the RAS algorithm in a form that is efficient both for computational efficiency and numerical stability. The standard statement of the algorithm may be found, for instance, in Schneider and Zenios (1990).

5. THE PARALLEL IMPLEMENTATIONS OF THE PRIMAL AND DUAL ALGORITHMS

The parallel computing platform that we used is a network of 16 Transputers installed in an Intel 286 personal computer operating under MS-DOS. This is probably one of the least costly parallel computing platforms available. The Transputers that we worked with are the T800 of Inmos/SGS Thompson, which have the following characteristics:

- each has four communication links which may send and receive data at 10 or 20 Mbits/sec.
- each has a 32 bit CPU with 4 Kbyte of "on-chip" memory and 1 or 4 Mbyte RAM.
- each operates at 20 Mhz frequency and has a rating of 10 MIPS (Million Instructions/second).
- each has a floating point processor rated at 1.5 Mflop (Million floating point operations per second).

A network of Transputers is configured by connecting their communication ports two by two. The only constraint is that each Transputer may be connected to at most four others. In order to configure the Transputers into a network we used a Linkputer, which may be programmed to obtain all possible configurations of up to 32 Transputers, with the restriction that a Transputer may be connected to at most four others. For the results presented in this paper we used 16 Transputers configured as a binary tree for the primal method and as a ring for the RAS algorithm. We used the EXPRESS library of Parasoft and the 3L Parallel Fortran 77 compiler as the development environment.

The parallel implementation of the projected gradient algorithm for the computation of the gradient is subdivided over the processors. Each processor p ($1 \leq p \leq NPROC$) computes a block (of columns) of ∇F^r (Step 1) that corresponds to $j \in J_p$, where J_p is a subset of the indices $j = 1, \ldots, m$. Each processor receives the data required for the computations at the beginning, and at each step it computes the components of the gradient. Then each processor computes $\frac{1}{n}\sum_{l=1}^{n} \nabla F_{lj}^r$ for $j \in J_p$ and $\frac{1}{m}\sum_{k \in J_p} \nabla F_{ik}^r$ (Step 2). Each processor in the tree receives the sums from its children and sends the partial sum to its father. The total is obtained at the root, which broadcasts it to all the processors.

The line search (Step 3) requires the evaluation of the objective function for various values of the step size α. A typical evaluation is the term

$$\sum_i \sum_j (g_{ij} - \alpha P_{ij}^r)\left(\ln \frac{(g_{ij}^r - \alpha P_{ij}^r)}{g_{ij}^0} - 1\right).$$

As in the previous steps, each processor p computes the part of the sum for indices $j \in J_p$. The communication and summation is done as in step 2.

The coarse grain parallel implementation of the RAS algorithm consists in the decomposition of the computations of steps 1 and 2 over the processors. In step 1, processor j computes S_j^r for $j \in J_p$, where the sets J_p form a partition of the columns $j = 1, 2, \ldots, m$ and have approximately the same cardinality. The communication of the partial results is done as follows: processors exchange results by successively sending and receiving information to and from their neighbors, in a ring network topology (Bertsekas and Tsitsiklis, 1989).

Substep 0: each processor p sends $S_{J_p}^r$ to processor $p+1$.

Substep 1: $(NPROC - 1)$ times each processor p receives $S_{J_{p'}}$ from processor $p-1$ and sends $S_{J_{p'}}^r$ to processor $p+1$.

The same approach is taken to partition the computation of the R_i^r in step 2 of the algorithm.

This is illustrated by the following example. Let the number of processors be 4 and the number of columns be 8. Each processor is assigned the computation of the S_j's for two columns. Processor 1 sends the values of S_1 and S_2, processor 2 sends the values of S_3 and S_4 and so on. Each processor then informs its neighbor of the values received. For this example, processor 1 would send to processor 2 the values of S_7 and S_8 that were received from processor 4, processor 2 sends the values of S_1 and S_2, and so on. This is illustrated in the figure below.

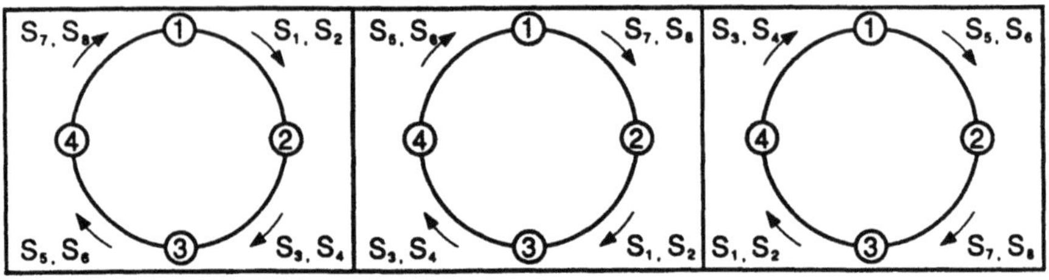

6. NUMERICAL RESULTS

We conducted our numerical experiments on problems with sizes not exceeding 315×315 in order to keep all the data in the RAM of each Transputer. The problems are generated randomly by the following procedure:

$\bar{g}_{ij}$ are randomly generated from a uniform distribution on $[1, 200]$.

$$O_i = \sum_{j=1}^{n} \bar{g}_{ij}, \quad \forall i; \qquad D_j = \sum_{i=1}^{m} \bar{g}_{ij}, \quad \forall j.$$

$g_{ij}^0 = \bar{g}_{ij}\,(1 \pm r_{ij})$, where r_{ij} are randomly generated from a uniform distribution on $[0, 0.1]$.

The increases in speed obtained with the primal method are reported in Table 1. The best increase is 11.55 and is obtained with 12 processors. The results obtained with the dual algorithm are reported in Table 2. Here the best increase is 5.03 using 16 processors.

These results indicate that the primal method benefits more from coarse grain parallelization. In addition, the convergence of the primal method is far superior to the dual method on ill conditioned examples such as that taken from Robillard and Stewart (1974). The data for the problem are

$$g^0 = \begin{bmatrix} 0.001 & 1 \\ 1 & 1 \end{bmatrix}, \qquad O = D = \begin{bmatrix} 2 \\ 2 \end{bmatrix}.$$

The exact solution is

$$x^* = \begin{bmatrix} \frac{2}{1+\sqrt{1000}} & \frac{2\sqrt{1000}}{1+\sqrt{1000}} \\ \frac{2\sqrt{1000}}{1+\sqrt{1000}} & \frac{2}{1+\sqrt{1000}} \end{bmatrix} = \begin{bmatrix} .06131\ldots & 1.93869\ldots \\ 1.93869\ldots & .06131\ldots \end{bmatrix}.$$

The RAS method produces the following solution after 150 iterations:

TABLE 1
Speedups obtained with the projected gradient algorithm.

Projected gradient method									
NPROC	2	3	4	6	8	10	12	14	16
Function + gradient evaluations	2	3	4	5.98	8	10	12	14	<u>16</u>
Gradient Projection	1.99	2.85	3.90	5.56	6.91	8.12	8.84	9.50	<u>9.95</u>
Line-Search	2	3	4	6	7.79	9.81	<u>11.85</u>	10.46	10.95
Total algorithm	2	3	4	6	7.69	9.63	<u>11.55</u>	10.46	10.98

TABLE 2
Speedups obtained with the RAS method.

Balancing method									
NPROC	2	3	4	6	8	10	12	14	16
Speedup	1.74	2.34	2.81	3.52	4.00	4.38	4.64	4.84	5.03

$$g^{(150)} = \begin{bmatrix} .0612 & 1.9387 \\ 1.9387 & .0613 \end{bmatrix},$$

while the projected gradient method produced the following solution after one (!) iteration:

$$g^{(1)} = \begin{bmatrix} .06131 & 1.93869 \\ 1.93869 & .06131 \end{bmatrix}.$$

A similar example is the following:

$$g^0 = \begin{bmatrix} 0.001 & .001 & 1 \\ 0.001 & 1 & 1 \\ 1 & 1 & 1 \end{bmatrix}, \qquad O = D = \begin{bmatrix} 3 \\ 3 \\ 3 \end{bmatrix}.$$

The corresponding primal and dual solutions are

$$\text{Projected gradient,} \quad g^{(16)} = \begin{bmatrix} 0.27053 & 0.02702 & 2.70245 \\ 0.02702 & 2.70245 & 0.27053 \\ 2.70245 & 0.27053 & 0.02702 \end{bmatrix},$$

$$\text{RAS algorithm,} \quad g^{(25)} = \begin{bmatrix} 0.1680 & 0.0214 & 2.8105 \\ 0.0207 & 2.6333 & 0.3460 \\ 2.6224 & 0.3338 & 0.0439 \end{bmatrix}.$$

A comparison of the sequential and parallelized algorithms for one 315×315 problem is given in the table below. The times are the average seconds per iteration.

Projected Gradient		RAS method	
Sequential	Parallel	Sequential	Parallel
15.96	1.38	2.09	0.42

Most of the computational time for the primal method is needed in the line search, 8.71 sec. out of 15.96 sec. The time required for the projection is 1.70 sec. and for the cost evaluation, 5.55 sec.

The ratio of the computational times, per iteration, of the parallelized algorithms is 3.28 in favor of the RAS algorithm. Hence, if the primal algorithm is to be competitive with the RAS method, it should converge to an acceptable solution in less than an average of 3.28 iterations. The preliminary tests given in this paper indicate that the primal algorithm converges faster than the RAS method for some ill conditioned examples. We note that the stopping criteria for the primal and dual methods are different. The pattern that we observe is that the RAS method obtains an objective value very close to that of the primal algorithm, but the values of the variables are relatively far from their optimal values.

7. CONCLUSION

We present in this paper, to our best knowledge, the first parallel computing implementations of a matrix balancing algorithm on an MIMD computing platform. The results obtained indicate excellent gains in speed for the primal algorithm, since the computing tasks are relatively "coarse grained" and well suited for this architecture. The gains in speed for the RAS algorithm may possibly be improved by an asynchronous implementation in which each processor does not wait to obtain the most "current" scaling factors before starting another balancing iteration. We intend to report on further work in this area in future articles.

REFERENCES

1. Bacharach, M., *"Biproportional matrices and input-output change"*, Cambridge University Press, 1970.
2. Bachem, A. and Korte, B., "On the RAS-Algorithm", *Computing*, 25, pp. 189–198, 1979.
3. Bertsekas, D.P. and Tsitsiklis, J.N., *"Parallel and distributed computation, numerical methods"*, Prentice Hall, Englewood Cliffs, New Jersey, 1989.
4. Bregman, L., "The relaxation method of finding the common point of convex sets and its application to the solution of problems in convex programming", *USSR Computational Math. and Mathematical Phys.*, 7, pp. 200–217, 1967.

5. Bregman, L., "Proof of the convergence of Sheleikhovskii's method for a problem with transportation constraints", *USSR Computational Math. and Mathematical Phys.*, 7, pp. 191–204, 1967.
6. Cottle, R.W., Duval, S.G. and Zikan, K., "A Lagrangean relaxation algorithm for the constrained matrix problem", *Naval Research Logistics Quarterly*, 33, pp. 55–76, 1986.
7. Drissi-Kaïtouni, O., "A projective method for bipartite networks and application to the matrix estimation and transportation problems", Publication #766, Centre for Research on Transportation, Université de Montréal, 1991.
8. Evans, A.W., "Some properties of trip distribution methods", *Transportation Research*, 4, pp. 19–36, 1970.
9. Evans, S.P. and Kirby, H.R., "A three-dimensional Furness procedure for calibrating gravity models", *Transportation Research*, 8, pp. 105–122, 1974.
10. Flynn, M.J., "Some computer organizations and their effectiveness", *IEEE Transactions on Computers*, C-21(9), pp. 948–960, 1972.
11. Furness, K.P., "Trip forecasting", *Unpublished paper cited by Evans and Kirby*, 1974.
12. Kruithof, J., "Calculation of telephone traffic", *De Ingenieur*, 52, pp. 15–25, 1937.
13. Lent, A., "A convergent algorithm for maximum entropy image restoration, with a medical X-ray application", *SPSE Conference Proceedings*, Toronto, Canada, 1976.
14. Pardalos, P.M., Phillips, A.T. and Rosen, J.B., *"Topics in parallel computing in mathematical programming"*, Department of Computer Science, The Pennsylvania State University, 1990.
15. Robillard, P. and Stewart, N.F., "Iterative numerical methods for trip distribution problems", *Transportation Research*, 8, pp. 575–582, 1974.
16. Schneider, M.H. and Zenios, S.A., "A comparative study of algorithms for matrix balancing", *Operations Research*, 38, 1990.
17. Wilson, A.G., *"Urban and Regional Models in Geography and Planning"*, Wiley, New York, 1974.
18. Zenios, S.A., "Parallel numerical optimization: current status and an annotated bibliography", *ORSA Journal on Computing*, vol. 1, No. 1, 1989.
19. Zenios, S.A. and Iu, S.L., "Vector and parallel computing for matrix balancing", *Annals of Operations Research*, 22, pp. 161–180, 1990.
20. Zenios, S.A., "Matrix balancing on a massively parallel connection machine", *ORSA Journal on Computing*, 2, pp. 112–125, 1990.

PART FOUR

The Computer and Econometric Studies

ANNA NAGURNEY AND JUNE DONG

Variational Inequalities for the Computation of Financial Equilibria in the Presence of Taxes and Price Controls

ABSTRACT. In this paper we develop a financial model of competitive sectors in the presence of policy interventions in the form of taxes and price ceilings. The model yields the equilibrium asset, liability, and financial instrument price pattern. First, the variational inequality formulation of the equilibrium conditions is derived and then utilized to obtain qualitative properties of the equilibrium pattern. We then propose a computational procedure and establish convergence results. The algorithm decomposes the large-scale problems into network subproblems of special structure, each of which can then be solved exactly (and simultaneously) in closed form. Numerical results are also presented to illustrate the algorithm's performance.

1. INTRODUCTION

In this paper we develop a framework for the formulation, analysis, and computation of competitive financial models in the presence of policy interventions. The policy interventions that are considered are taxes and price controls.

Financial theory since the seminal work of Markowitz (1959) and Sharpe (1970) has been principally concerned with the portfolio optimization problem facing a single sector. Here, by contrast, we focus on the competitive equilibrium problem in which there are multiple sectors, each with its particular optimization problem. We assume that each sector seeks to minimize its risk while maximizing its net return in the presence of both taxes and price ceilings. Under the assumption of perfect competition, each sector in the economy takes the instrument prices as given and then determines its optimal composition of both assets and liabilities subject to an accounting identity. In the model, the instrument prices serve as market signals which, in turn, reflect the economic market conditions that state that if an instrument price at equilibrium lies within the bounds, then the market for the financial instrument must clear.

The theoretical developments in this paper are done using variational inequality theory. We note that variational inequality theory has been used to study a plethora of problems, including oligopolistic market equilibrium problems (cf. Gabay and Moulin, 1980; Dafermos and Nagurney, 1987), spatial price equilibrium problems (cf. Florian and Los, 1982; and Dafermos and Nagurney, 1984), and general economic equilibrium problems (cf. Border, 1985; Dafermos, 1990; and Dafermos and Zhao, 1991). In addition, variational inequality theory has been used recently to investigate the effects of policy interventions in the form of price controls in commodity markets

D. A. Belsley (ed.), Computational Techniques for Econometrics and Economic Analysis, 189–205.

(cf. Nagurney and Zhao, 1991). These, however, have been partial equilibrium models in which only a subset of agents/commodities has been treated. In this paper we demonstrate that variational inequality theory can also be used to study policy interventions in *general* equilibrium problems, in particular, in general financial equilibrium problems.

The paper is organized as follows. In Section 2 we introduce the model, consisting of multiple sectors and multiple financial instruments that can be held as assets and/or as liabilities in the presence of taxes and price ceilings. The model postulates the behavior of the sectors and, in equilibrium, yields the competitive asset and liability holdings as well as the instrument prices. The variational inequality formulation of the equilibrium conditions is derived and then used to study the qualitative properties. We first show that a solution is guaranteed to exist and then establish uniqueness of the equilibrium asset and liability pattern.

In Section 3 we propose an algorithm for the computation of the equilibrium pattern. The algorithm is the modified projection method of Korpelevich (1977), which is shown to converge for our model. The notable feature of this decomposition algorithm is that it resolves the large-scale financial problem into simple network subproblems of special structure, each of which can be solved simultaneously and exactly in closed form using exact equilibration algorithms (cf. Eydeland and Nagurney, 1989). In Section 4 we conduct numerical experiments with the algorithm on a variety of examples. In Section 5 we summarize the results and present our conclusions.

2. THE MODEL OF COMPETITIVE FINANCIAL EQUILIBRIUM WITH POLICY INTERVENTIONS

In this section we develop a model of competitive financial equilibrium that permits the incorporation of policy interventions in the form of taxes and price controls. The behavior of the sectors is stated along with the market conditions for the financial instruments. The equilibrium conditions are then formulated as a variational inequality problem. Finally, the qualitative properties of existence and uniqueness are discussed.

We consider an economy consisting of m sectors, with typical sector i, and with n instruments, with typical instrument j. The volume of instrument j held in sector i's portfolio as an asset is denoted by x_{ij}, and the volume of instrument j held in sector i's portfolio as a liability by y_{ij}. The assets in sector i's portfolio are grouped into a column vector $x_i \in R^n$, and the liabilities are grouped into the column vector $y_i \in R^n$. We further group the sector asset vectors into the vector $x \in R^{mn}$ and the sector liability vectors into the vector $y \in R^{mn}$.

Assume that each sector's disutility can be defined through its assessment of risk with respect to its portfolio composition minus its total expected net yield plus its total tax payment. Each sector's risk is represented by a variance-covariance matrix denoting the sector's assessment of the standard deviation of prices for each instrument. The $2n \times 2n$ variance-covariance matrix associated with sector i's assets and liabilities is denoted by Q^i.

In this model we also assume that the total volume of each balance sheet side is exogenous. Moreover, under the assumption of perfect competition, each sector will behave as if it has no influence on instrument prices or on the behavior of the other sectors. Let r_j denote the price of instrument j, and group the instruments into the vector $r \in R^n$.

We now describe the policy interventions. First, denote the tax rate levied on sector i's net yield on financial instrument j as τ_{ij} and group the tax rates into the vector $\tau \in R^{mn}$. We assume that the tax rates lie in the interval $0 \leq \tau < 1$. Note therefore that the government in this model has the flexibility of levying a distinct tax rate across both sectors and instruments. Further, denote the price ceiling associated with instrument j by $\hat{r}_j$, and group the ceilings into a vector $\hat{r} \in R^n$. Note that ceilings have been imposed on variables in other models, in particular on commodity prices in spatial price equilibrium problems (cf. Thore (1986) and Nagurney and Zhao (1991)).

Each sector i, then, seeks to determine its optimal composition of instruments held as assets and as liabilities so as to minimize risk while, at the same time, maximizing expected net yield subject to the tax rate structure. Each sector, under the assumption of perfect competition, takes the instrument prices as given. The portfolio optimization problem for sector i can be stated mathematically as

$$\text{Minimize} \quad \begin{pmatrix} x_i \\ y_i \end{pmatrix}^T Q^i \begin{pmatrix} x_i \\ y_i \end{pmatrix} - \sum_{j=1}^{n} (1 - \tau_{ij}) r_j (x_{ij} - y_{ij})$$

subject to

$$\sum_{j=1}^{n} x_{ij} = s_i, \tag{1}$$

$$\sum_{j=1}^{n} y_{ij} = s_i,$$

$$x_{ij} \geq 0, \quad y_{ij} \geq 0; \quad j = 1, \ldots, n. \tag{2}$$

Constraints (1) represent the accounting identities that require that the accounts for sector i must balance, where s_i is the total financial volume held by sector i. Constraint (2) represents the nonnegativity assumption. We let P_i denote the closed convex set of (x_i, y_i) satisfying constraints (1) and (2).

Since Q^i is a variance-covariance matrix, we will assume it is positive-definite, and therefore the objective function for each sector is strictly convex. Thus, the necessary and sufficient conditions for an optimal portfolio are that $(x_i, y_i) \in P_i$ must satisfy the following system of inequalities and equalities:

For each instrument $j = 1, \ldots, n$,

$$2Q^i_{(11)j}{}^T \cdot x_i + 2Q^i_{(21)j}{}^T \cdot y_i - (1 - \tau_{ij}) r_j - \mu^1_i \geq 0,$$

$$2{Q^i_{(22)j}}^T \cdot y_i + 2{Q^i_{(12)j}}^T \cdot x_i + (1-\tau_{ij})r_j - \mu^2_i \geq 0,$$

$$x_{ij} \cdot (2{Q^i_{(11)j}}^T \cdot x_i + 2{Q^i_{(21)j}}^T \cdot y_i - (1-\tau_{ij})r_j - \mu^1_i) = 0, \tag{3}$$

$$y_{ij} \cdot (2{Q^i_{(22)j}}^T \cdot y_i + 2{Q^i_{(12)j}}^T \cdot x_i + (1-\tau_{ij})r_j - \mu^2_i) = 0,$$

where the symmetric matrix Q^i has been partitioned as $Q^i = \begin{pmatrix} Q^i_{11} & Q^i_{12} \\ Q^i_{21} & Q^i_{22} \end{pmatrix}$ and $Q^i_{(\alpha\beta)j}$ denotes the j-th column of $Q^i_{(\alpha\beta)}$, with $\alpha = 1,2; \beta = 1,2$. The terms μ^1_i and μ^2_i are the Lagrange multipliers of constraints (1). A similar set of inequalities and equalities will hold for each of the m sectors.

We now describe the inequalities governing the instrument prices in the economy. Note that the prices provide feedback to the sectors through the objective function. We assume that there is free disposal and, hence, the instrument prices will be nonnegative. Mathematically, the economic system conditions are thus:

For each instrument $j = 1, \ldots, n$,

$$\sum_{i=1}^{m} (1-\tau_{ij})(x_{ij} - y_{ij}) \begin{cases} \leq 0, & \text{if } r_j = \hat{r}_j \\ = 0, & \text{if } r_j > 0 \quad \text{and } r_j < \hat{r}_j \\ \geq 0, & \text{if } r_j = 0. \end{cases} \tag{4}$$

Therefore, if there is an effective excess supply of any instrument in the economy, its price must be zero; if an instrument's price is positive (but not at the ceiling), the market for that instrument must clear; and, if there is an effective excess demand for an instrument in the economy, its price must be at the ceiling.

Let $S \equiv \{r | 0 \leq r \leq \hat{r}\}$, and $K \equiv \prod_{i=1}^{m} P_i \times S$, and combine the above sector and market inequalities and equalities to obtain

Definition 1.

A vector $(x, y, r) \in K$ is an equilibrium point of the competitive financial model with policy interventions developed above if and only if it satisfies the system of equalities and inequalities (3) and (4) for all sectors $i = 1, \ldots, m$ and for all instruments $j = 1, \ldots, n$ simultaneously.

We now derive the variational inequality formulation of the equilibrium conditions of the above model in the subsequent theorem.

Theorem 1.

A vector (x, y, r) of assets and liabilities of the sectors and instrument prices is a competitive financial equilibrium with policy interventions if and only if it satisfies the variational inequality problem:

Find $(x, y, r) \in K$ satisfying

$$\sum_{i=1}^{m}\sum_{j=1}^{n} \left[2({Q^i_{(11)j}}^T \cdot x_i + {Q^i_{(21)j}}^T \cdot y_i) - (1-\tau_{ij})r_j\right] \times \left[x'_{ij} - x_{ij}\right]$$

$$+\sum_{i=1}^{m}\sum_{j=1}^{n}\left[2({Q^i_{(22)j}}^T\cdot y_i+{Q^i_{(12)j}}^T\cdot x_i)+(1-\tau_{ij})r_j\right]\times[y'_{ij}-y_{ij}]$$

$$+\sum_{j=1}^{n}\left[\sum_{i=1}^{m}(1-\tau_{ij})\,(x_{ij}-y_{ij})\right]\times[r'_j-r_j]\geq 0, \tag{5}$$

for all $(x', y', r') \in K$.

Proof: Assume that $(x, y, r) \in K$ is an equilibrium point. Then inequalities (3) and (4) hold for all i and j. Hence, one has

$$\sum_{j=1}^{n}\left[2({Q^i_{(11)j}}^T\cdot x_i+{Q^i_{(21)j}}^T\cdot y_i)-(1-\tau_{ij})r_j-\mu^1_i\right]\times\left[x'_{ij}-x_{ij}\right]\geq 0,$$

from which it follows, after applying constraint (1), that

$$\sum_{j=1}^{n}\left[2({Q^i_{(11)j}}^T\cdot x_i+{Q^i_{(21)j}}^T\cdot y_i)-(1-\tau_{ij})r_j\right]\times\left[x'_{ij}-x_{ij}\right]\geq 0. \tag{6}$$

Similarly, one can obtain

$$\sum_{j=1}^{n}\left[2({Q^i_{(22)j}}^T\cdot y_i+{Q^i_{(12)j}}^T\cdot x_i)+(1-\tau_{ij})r_j\right]\times\left[y'_{ij}-y_{ij}\right]\geq 0. \tag{7}$$

Summing inequalities (6) and (7) over i, one concludes that for $(x, y) \in \prod_{i=1}^{m} P_i$,

$$\sum_{i=1}^{m}\sum_{j=1}^{n}\left[2({Q^i_{(11)j}}^T\cdot x_i+{Q^i_{(21)j}}^T\cdot y_i)-(1-\tau_{ij})r_j\right]\times\left[x'_{ij}-x_{ij}\right]$$

$$+\sum_{i=1}^{m}\sum_{j=1}^{n}\left[2({Q^i_{(22)j}}^T\cdot y_i+{Q^i_{(12)j}}^T\cdot x_i)+(1-\tau_{ij})r_j\right]\times\left[y'_{ij}-y_{ij}\right]\geq 0 \tag{8}$$

for all $(x', y') \in \prod_{i=1}^{m} P_i$.

From inequality (4), one can conclude that $0 \leq r_j \leq \hat{r}_j$ must satisfy

$$\sum_{i=1}^{m}(1-\tau_{ij})(x_{ij}-y_{ij})\times(r'_j-r_j)\geq 0 \tag{9}$$

for all $0 \leq r'_j \leq \hat{r}_j$, and, therefore, $r \in S$ must satisfy

$$\sum_{j=1}^{n}\sum_{i=1}^{m}(1-\tau_{ij})(x_{ij}-y_{ij})\times(r'_j-r_j)\geq 0 \tag{10}$$

for all $r' \in S$. Summing inequalities (8) and (10) produces the variational inequality (5).

We now establish that a solution to variational inequality (5) will also satisfy equilibrium conditions (3) and (4). If $(x, y, r) \in K$ is a solution of variational inequality (5) and if one lets $x_i' = x_i, y_i' = y_i$, for all i, one obtains

$$\sum_{j=1}^{n} \left[\sum_{i=1}^{m} (1 - \tau_{ij})(x_{ij} - y_{ij}) \right] \times \left[r_j' - r_j \right] \geq 0, \tag{11}$$

for all $r' \in S$, which implies conditions (4).

Finally, let $r_j' = r_j$, for all j, in which case substitution into (5) yields

$$\sum_{i=1}^{m} \sum_{j=1}^{n} \left[2({Q^i_{(11)j}}^T \cdot x_i + {Q^i_{(21)j}}^T \cdot y_i) - (1 - \tau_{ij}) r_j \right] \times \left[x_{ij}' - x_{ij} \right]$$

$$+ \sum_{i=1}^{m} \sum_{j=1}^{n} \left[2({Q^i_{(22)j}}^T \cdot x_i + {Q^i_{(12)j}}^T \cdot y_i) + (1 - \tau_{ij}) r_j \right] \times \left[y_{ij}' - y_{ij} \right] \geq 0 \tag{12}$$

for all $(x', y') \in \prod_{i=1}^{m} P_i$, which implies (3). The proof is complete.

We now address the qualitative properties of the equilibrium pattern through the study of variational inequality (5). Observe that the feasible set K is compact and that the function that enters the variational inequality (5) is continuous. It follows from the standard theory of variational inequalities (cf. Kinderlehrer and Stampacchia (1980)) that the solution (x, y, r) to (5) is guaranteed to exist. The uniqueness result is given in the following theorem.

Theorem 2. The equilibrium asset and liability pattern (x, y) is unique.

Proof: Suppose that $(x^1, y^1, r^1) \in K$ and $(x^2, y^2, r^2) \in K$ both satisfy variational inequality (5), so

$$\sum_{i=1}^{m} \sum_{j=1}^{n} \left[2({Q^i_{(11)j}}^T \cdot x_i^1 + {Q^i_{(21)j}}^T \cdot y_i^1) - (1 - \tau_{ij}) r_j^1 \right] \times \left[x_{ij}' - x_{ij}^1 \right]$$

$$+ \sum_{i=1}^{m} \sum_{j=1}^{n} \left[2({Q^i_{(22)j}}^T \cdot y_i^1 + {Q^i_{(12)j}}^T \cdot x_i^1) + (1 - \tau_{ij}) r_j^1 \right] \times \left[y_{ij}' - y_{ij}^1 \right]$$

$$+ \sum_{j=1}^{n} \left[\sum_{i=1}^{m} (1 - \tau_{ij})(x_{ij}^1 - y_{ij}^1) \right] \times \left[r_j' - r_j^1 \right] \geq 0 \tag{13}$$

for all $(x', y,' r') \in K$, and

$$\sum_{i=1}^{m}\sum_{j=1}^{n}\left[2(Q^{i}_{(11)j}{}^{T}\cdot x^2_i + Q^{i}_{(21)j}{}^{T}\cdot y^2_i) - (1-\tau_{ij})r^2_j\right]\times\left[x'_{ij}-x^2_{ij}\right]$$

$$+\sum_{i=1}^{m}\sum_{j=1}^{n}\left[2(Q^{i}_{(22)j}{}^{T}\cdot y^2_i + Q^{i}_{(12)j}{}^{T}\cdot x^2_i) + (1-\tau_{ij})r^2_j\right]\times\left[y'_{ij}-y^2_{ij}\right]$$

$$+\sum_{j=1}^{n}\left[\sum_{i=1}^{m}(1-\tau_{ij})(x^2_{ij}-y^2_{ij})\right]\times\left[r'_j - r^2_j\right]\geq 0 \tag{14}$$

for all $(x', y', r') \in K$.

Setting $(x', y', r') = (x^2, y^2, r^2)$ in (13) and $(x', y', r') = (x^1, y^1, r^1)$ in (14) and adding the results yields, after simplification,

$$\sum_{i=1}^{m}\sum_{j=1}^{n}\left[2(Q^{i}_{(11)j}{}^{T}\cdot(x^1_i - x^2_i) + Q^{i}_{(21)j}{}^{T}\cdot(y^1_i - y^2_i))\right]\times\left[x^2_{ij}-x^1_{ij}\right]$$

$$+\sum_{i=1}^{m}\sum_{j=1}^{n}\left[2(Q^{i}_{(22)j}{}^{T}\cdot(y^1_i - y^2_i) + Q^{i}_{(12)j}{}^{T}\cdot(x^1_i - x^2_i))\right]\times\left[y^2_{ij}-y^1_{ij}\right]\geq 0. \tag{15}$$

But since each Q^i; $(i = 1, \ldots, m)$ is positive-definite, the left-hand side of (15) must be nonpositive and, hence, we can conclude that $x^1 = x^2$ and $y^1 = y^2$. The proof is complete.

In the special case in which there are no taxes or price ceilings imposed, the above model collapses to the model developed in Nagurney, Dong, and Hughes (1992). In this case condition (4) simplifies as follows. Since $\tau_{ij} = 0$ for all i and j, and $\hat{r}_j$ is effectively set at infinity for all j, only the equality and the second inequality would apply in conditions (4). In addition, in the model without policies, the set S would no longer be bounded, and, hence, the feasible set K would no longer be compact. Hence, another existence proof would be required.

3. THE ALGORITHM

In this Section we describe a decomposition algorithm for the computation of the solution to variational inequality (5) that governs the general competitive financial equilibrium model with taxes and price controls developed in Section 2. The algorithm resolves the large-scale problems into simpler network subproblems, each of which can then be solved explicitly and in closed form using exact equilibration algorithms (cf. Eydeland and Nagurney (1989), and the references therein). The algorithm that we propose is the modified projection method of Korpelevich (1977). We first state the algorithm in the context of the financial model and then prove convergence.

The Algorithm

Step 0: Initialization:

Set $(x^0, y^0, r^0) \in K$. Let $k = 1$. Let α be a positive scalar.

Step 1: Compute $(\bar{x}^k, \bar{y}^k, \bar{r}^k)$ by Solving

$$\sum_{i=1}^{m} \sum_{j=1}^{n} \Big[\bar{x}_{ij}^k + \alpha(2({Q_{(11)j}^i}^T \cdot x_i^{k-1} + {Q_{(21)j}^i}^T \cdot y_i^{k-1})$$

$$-(1-\tau_{ij}) r_j^{k-1}) - x_{ij}^{k-1}\Big] \cdot \big[x'_{ij} - \bar{x}_{ij}^k\big]$$

$$+\sum_{i=1}^{m} \sum_{j=1}^{n} \Big[\bar{y}_{ij}^k + \alpha(2{Q_{(22)j}^i}^T \cdot y_i^{k-1} + {Q_{(12)j}^i}^T \cdot x_i^{k-1})$$

$$+(1-\tau_{ij}) r_j^{k-1}) - y_{ij}^{k-1}\Big] \cdot \big[y'_{ij} - \bar{y}_{ij}^k\big]$$

$$+\sum_{j=1}^{n} \left[\bar{r}_j^k + \alpha \sum_{i=1}^{m} (1-\tau_{ij})(x_{ij}^{k-1} - y_{ij}^{k-1}) - r_j^{k-1}\right]$$

$$\cdot \big[r'_j - \bar{r}_j^k\big] \geq 0, \forall (x', y', r') \in K. \tag{16}$$

Step 2: Compute (x^k, y^k, r^k) by Solving

$$\sum_{i=1}^{m} \sum_{j=1}^{n} \Big[x_{ij}^k + \alpha(2({Q_{(11)j}^i}^T \cdot \bar{x}_i^k + {Q_{(21)j}^i}^T \cdot \bar{y}_i^k)$$

$$-(1-\tau_{ij}) \bar{r}_j^k) - x_{ij}^{k-1}\Big] \cdot \big[x'_{ij} - x_{ij}^k\big]$$

$$+\sum_{i=1}^{m} \sum_{j=1}^{n} \Big[y_{ij}^k + \alpha(2({Q_{(22)j}^i}^T \cdot \bar{y}_i^k + {Q_{(12)j}^i}^T \cdot \bar{x}_i^k)$$

$$+(1-\tau_{ij}) \bar{r}_j^k) - y_{ij}^{k-1}\Big] \cdot \big[y'_{ij} - y_{ij}^k\big]$$

$$+\sum_{j=1}^{n} \left[r_j^k + \alpha \sum_{i=1}^{m} (1-\tau_{ij})(\bar{x}_{ij}^{k-1} - \bar{y}_{ij}^{k-1}) - r_j^{k-1}\right]$$

$$\cdot \big[r'_j - r_j^k\big] \geq 0, \forall (x', y', r') \in K. \tag{17}$$

Convergence Verification:

If $\max_{ij} |x_{ij}^k - x_{ij}^{k-1}| \leq \epsilon$, $\max_{ij} |y_{ij}^k - y_{ij}^{k-1}| \leq \epsilon$, $\max_{j} |r_j^k - r_j^{k-1}| \leq \epsilon$, with ϵ some positive preselected tolerance, then stop; else, set $k = k + 1$, and go to Step 1.

We now give an interpretation of the above algorithm as an adjustment process. In (16) each sector i at each time period k receives instrument price signals r^{k-1} and determines its optimal asset and liability pattern $\bar{x}_i^k, \bar{y}_i^k$. At the same time, the system determines the prices $\bar{r}^k$ in response to the difference of the total effective volume of each instrument held as an asset minus the total effective volume held as a liability at time period $k - 1$. The agents and the system then improve upon their approximations through the solution of (17). The process continues until stability is reached; that is, until the asset and liability volumes and the instrument prices change negligibly between time periods.

Observe now that both (16) and (17) are equivalent to optimization problems, in particular to quadratic programming problems of the form

$$\text{Minimize}_{X \in K} \quad X^T \cdot X + h^T \cdot X, \tag{18}$$

where $X \equiv \{(x, y, r) \in R^{2mn+n}\}$ and $h \in R^{2mn+n}$ consists of the fixed linear terms in the inequality subproblems (16) and (17). Moreover, problem (18) is separable in x, y and r, and, in view of the feasible set, has the network structure depicted in Figure 1.

Convergence of the algorithm follows (cf. Korpelevich, 1977) under the assumption that the function F that enters the variational inequality is monotone and Lipschitz continuous, where $0 < \alpha < \frac{1}{L}$ and L is the Lipschitz constant. We now prove that these conditions are always satisfied. We first establish monotonicity.

Let $(x^1, y^1, r^1) \in K$ and $(x^2, y^2, r^2) \in K$. In evaluating monotonicity, we must show that

$$\sum_{i=1}^{m} \sum_{j=1}^{n} \Big[2({Q^i_{(11)j}}^T \cdot (x_i^1 - x_i^2) + {Q^i_{(21)j}}^T \cdot (y_i^1 - y_i^2))$$

$$-((1 - \tau_{ij})r_j^1 - (1 - \tau_{ij})r_j^2)\Big] \times (x_{ij}^1 - x_{ij}^2)$$

$$+\sum_{i=1}^{m} \sum_{j=1}^{n} \Big[2({Q^i_{(22)j}}^T \cdot (y_i^1 - y_i^2) + {Q_{(12)j}}^T \cdot (x_i^1 - x_i^2))$$

$$+((1 - \tau_{ij})r_j^1 - (1 - \tau_{ij})r_j^2)\Big] \times (y_{ij}^1 - y_{ij}^2)$$

$$+\sum_{j=1}^{n} \left[\sum_{i=1}^{m}(1 - \tau_{ij})(x_{ij}^1 - y_{ij}^1) - \sum_{i=1}^{m}(1 - \tau_{ij})(x_{ij}^2 - y_{ij}^2)\right] \times \left[r_j^1 - r_j^2\right] \geq 0, \tag{19}$$

for all $(x^1, y^1, r^1) \in K$, $(x^2, y^2, r^2) \in K$. After some algebra, the left-hand side of (19) reduces to

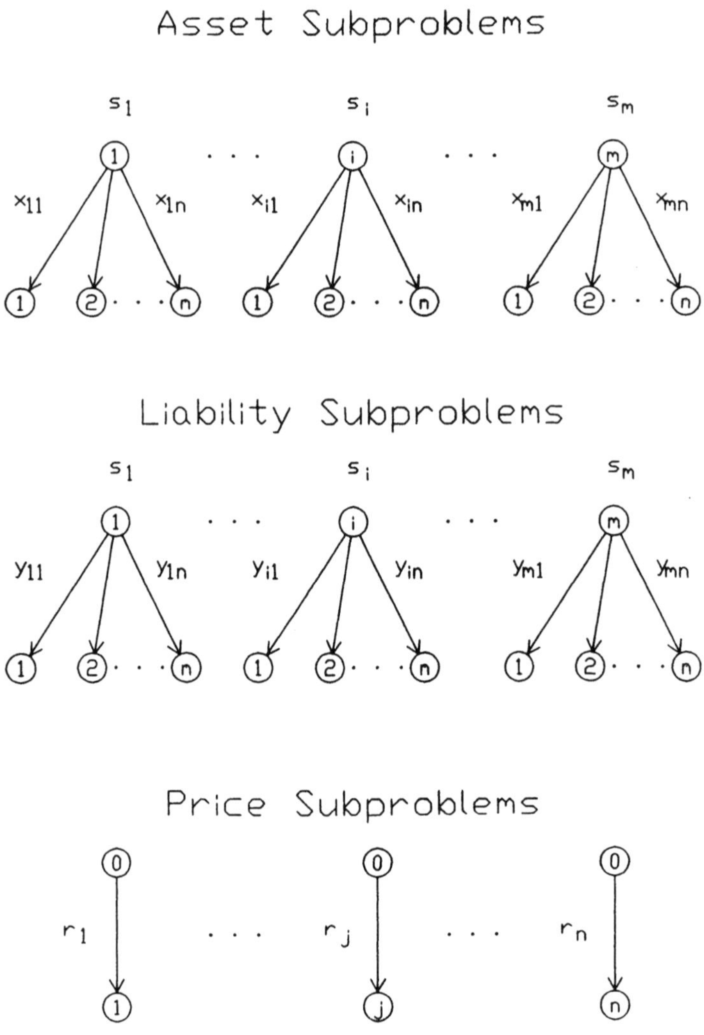

Fig. 1. Parallel network structure of the variational inequality subproblems.

$$\sum_{i=1}^{m}\sum_{j=1}^{n}\left[2(Q^{i}_{(11)j}{}^{T}\cdot(x^{1}_{i}-x^{2}_{i})+Q^{i}_{(21)j}{}^{T}\cdot(y^{1}_{i}-y^{2}_{i}))\right]\times(x^{1}_{ij}-x^{2}_{ij})$$

$$+\sum_{i=1}^{m}\sum_{j=1}^{n}\left[2(Q^{i}_{(22)j}{}^{T}\cdot(y^{1}_{i}-y^{2}_{i})+Q_{(12)j}{}^{T}\cdot(x^{1}_{i}-x^{2}_{i}))\right]\times(y^{1}_{ij}-y^{2}_{ij}),$$

which is clearly nonnegative since Q^{i} has been assumed to be positive-definite. We have thus established:

Lemma 1. The function that enters the variational inequality (5) is monotone.

We now investigate Lipschitz continuity in the following lemma.

Lemma 2. The function $F(x, y, r)$ that enters variational inequality (5) is Lipschitz continuous; that is, for all (x^1, y^1, r^1), $(x^2, y^2, r^2) \in K$,

$$\|F(x^1, y^1, r^1) - F(x^2, y^2, r^2)\|_2 \leq L\|(x^1, y^1, r^1) - (x^2, y^2, r^2)\|_2, \tag{20}$$

with Lipschitz constant $L > 0$.

Proof: $F(x, y, r)$ can be represented as

$$F(x, y, r) = C \begin{pmatrix} x \\ y \\ r \end{pmatrix}, \tag{21}$$

where C is of the form

$$\begin{pmatrix} Q & TB \\ -(TB)^T & 0 \end{pmatrix} \tag{22}$$

and

$$Q = \begin{pmatrix} Q_{11}^1 & & Q_{21}^1 & & \\ & \ddots & & \ddots & \\ & & Q_{11}^m & & Q_{21}^m \\ Q_{12}^1 & & & Q_{22}^1 & \\ & \ddots & & \ddots & \\ & & Q_{12}^m & & Q_{22}^m \end{pmatrix}, \tag{23}$$

$$T = \begin{pmatrix} 1-\tau_{11} & & & & & \\ & \ddots & & & & \\ & & 1-\tau_{mn} & & & \\ & & & 1-\tau_{11} & & \\ & & & & \ddots & \\ & & & & & 1-\tau_{mn} \end{pmatrix} \quad B = \begin{pmatrix} -I \\ \vdots \\ -I \\ I \\ \vdots \\ I \end{pmatrix}_{2mn \times n}. \tag{24}$$

$$\|F(x^1, y^1, r^1) - F(x^2, y^2, r^2)\|_2^2 = \begin{pmatrix} x^1 - x^2 \\ y^1 - y^2 \\ r^1 - r^2 \end{pmatrix}^T C^T C \begin{pmatrix} x^1 - x^2 \\ y^1 - y^2 \\ r^1 - r^2 \end{pmatrix} \tag{25}$$

and

$$\|(x^1, y^1, r^1) - (x^2, y^2, r^2)\|_2^2 = \begin{pmatrix} x^1 - x^2 \\ y^1 - y^2 \\ r^1 - r^2 \end{pmatrix}^T \begin{pmatrix} x^1 - x^2 \\ y^1 - y^2 \\ r^1 - r^2 \end{pmatrix}. \tag{26}$$

Since C^TC is a symmetric matrix, if we let

$$L^2 = \max_{1 \leq l \leq 2mn+n} \sum_{k=1}^{2mn+n} |c_{lk}| + 1, \tag{27}$$

where c_{lk} is the (l, k)th element of C^TC, then $(L^2I - C^TC)$ is a symmetric positive-definite matrix, and, therefore, $(C^TC - L^2I)$ is a negative-definite matrix. Hence,

$$\|F(x^1, y^1, r^1) - F(x^2, y^2, r^2)\|_2^2 - L^2\|(x^1, y^1, r^1) - (x^2, y^2, r^2)\|_2^2$$

$$= \begin{pmatrix} x^1 - x^2 \\ y^1 - y^2 \\ r^1 - r^2 \end{pmatrix}^T (C^TC - L^2I) \begin{pmatrix} x^1 - x^2 \\ y^1 - y^2 \\ r^1 - r^2 \end{pmatrix} \leq 0. \tag{28}$$

Consequently,

$$\|F(x^1, y^1, r^1) - F(x^2, y^2, r^2)\|_2 \leq L\|(x^1, y^1, r^1) - (x^2, y^2, r^2)\|_2, \tag{29}$$

with $L < 0$. The proof is complete.

Combining Lemmas 1 and 2 we obtain

Theorem 3. The decomposition algorithm converges to the equilibrium asset, liability, and price pattern (x, y, r) satisfying variational inequality (5).

4. NUMERICAL RESULTS

In this section we consider the numerical solution of the financial equilibrium model with policy interventions introduced in Section 2. We emphasize that the model is designed with empirical applications in mind. For example, the framework that has been developed fits well with flow-of-funds accounts data that are collected quarterly or annually to provide snapshots of the financial side of the economy. In the case of the United States, the data sets are maintained by the Federal Reserve Board of Governors. For an introduction to flow of funds accounts, see Cohen (1987), and for a recent algorithmic approach to the balancing of these accounts, see Hughes and Nagurney (1991).

We now present several examples with a variety of tax and price control scenarios. We consider an economy with two sectors and three financial instruments. Here we assume that the "size" of each sector s_i is given by $s_1 = 1$, and $s_2 = 2$. Each sector realizes that the future values of its portfolio are random variables that can be described by their expected values and variances and believes that the mean of these expected values is equal to the current value. The variance-covariance matrices of the two sectors are:

$$Q^1 = \begin{pmatrix} 1 & .15 & .3 & -.2 & -.1 & 0 \\ .15 & 1 & .1 & -.1 & -.2 & 0 \\ .3 & .1 & 1 & -.3 & 0 & -.1 \\ -.2 & -.1 & -.3 & 1 & 0 & .3 \\ -.1 & -.2 & 0 & 0 & 1 & .2 \\ 0 & 0 & -.1 & .3 & .2 & 1 \end{pmatrix}$$

and

$$Q^2 = \begin{pmatrix} 1 & .4 & .3 & -.1 & -.1 & 0 \\ .4 & 1 & .5 & 0 & -.05 & 0 \\ .3 & .5 & 1 & 0 & 0 & -.1 \\ -.1 & 0 & 0 & 1 & .5 & 0 \\ -.1 & -.05 & 0 & .5 & 1 & .2 \\ 0 & 0 & -.1 & 0 & .2 & 1 \end{pmatrix}.$$

Note that the terms in the blocks Q^1_{12}, Q^1_{21}, Q^2_{12}, Q^2_{21}, are not positive, since the returns flowing in from an asset item must be negatively correlated with the interest expenses flowing out into the portfolio's liabilities. (For details see Francis and Archer (1979).)

We now use the above data to construct the examples. The algorithm was coded in FORTRAN, compiled using the FORTVS compiler, optimization level 3, and the numerical runs were done on an IBM 3090/600J. For each example, the variables were initialized as follows: $r_j^0 = 1$, for all j, $x_{ij} = \frac{s_i}{n}$, for all j, $y_{ij} = \frac{s_i}{n}$, for all j. The α parameter was set to .35. The convergence tolerance ϵ was set to 10^{-3}.

In the first example, we set the taxes $\tau = 0$, for all sectors and instruments and the price control ceilings $\hat{r}$ to 2 for all instruments. The numerical results for the first example are

Results for Example 1
Equilibrium Prices:

$$r_1 = .91404 \quad r_2 = .94535 \quad r_3 = 1.14058$$

Equilibrium Asset Holdings:

$$x_{11} = .28736 \quad x_{12} = .40063 \quad x_{13} = .31200$$

$$x_{21} = .75644 \quad x_{22} = .56740 \quad x_{23} = .67616$$

Equilibrium Liability Holdings:

$$y_{11} = .32035 \quad y_{12} = .51047 \quad y_{13} = .16917$$

$$y_{21} = .72447 \quad y_{22} = .45723 \quad y_{23} = .81830.$$

The algorithm converged in 17 iterations and required 3.62 miliseconds of CPU time for convergence, not including input/output time. Note that in this example, the solution is one in which the policies, in essence, have no effect. Hence, this algorithm may also be used to compute solutions to financial models in the absence of taxes and price controls, provided that the taxes are set to zero and the price ceilings are set at a high enough level. The resulting model is then a special case of our more general one.

In the second example, we kept the taxes at zero but now tightened the price ceilings to .5 for each instrument. The numerical results for the second example are

Results for Example 2
Equilibrium Prices:

$$r_1 = .27083 \quad r_2 = .30192 \quad r_3 = .49716$$

Equilibrium Asset Holdings:

$$x_{11} = .28730 \quad x_{12} = .40043 \quad x_{13} = .31227$$

$$x_{21} = .75653 \quad x_{22} = .56752 \quad x_{23} = .67595$$

Equilibrium Liability Holdings:

$$y_{11} = .32005 \quad y_{12} = .51074 \quad y_{13} = .16920$$

$$y_{21} = .72464 \quad y_{22} = .45708 \quad y_{23} = .81828.$$

The algorithm converged in 18 iterations and required 3.82 miliseconds of CPU time for convergence. Note that in this example, the equilibrium prices all lie within the tighter bounds. In particular, the price of instrument 3 is approximately at its upper bound of .5

In the third example, we raised the tax rate from zero to .15 for all sectors and instruments and kept the instrument price ceiling at .5, as in Example 2. The numerical results for the third example are

Results for Example 3
Equilibrium Prices:

$$r_1 = .23256 \quad r_2 = .26871 \quad r_3 = .49995$$

Equilibrium Asset Holdings:

$$x_{11} = .28726 \quad x_{12} = .40035 \quad x_{13} = .31239$$

$x_{21} = .75663 \quad x_{22} = .56777 \quad x_{23} = .67560$

Equilibrium Liability Holdings:

$y_{11} = .31965 \quad y_{12} = .51098 \quad y_{13} = .16938$

$y_{21} = .72460 \quad y_{22} = .45680 \quad y_{23} = .81860.$

The algorithm converged in 19 iterations and required 4.04 miliseconds of CPU time for convergence.

In the fourth example, we kept the price ceilings at .5 but increased the tax rate from .15 to .30. The numerical results for this example are

Results for Example 4
Equilibrium Prices:

$r_1 = .17990 \quad r_2 = .22313 \quad r_3 = .5000.$

Equilibrium Asset Holdings:

$x_{11} = .28782 \quad x_{12} = .40104 \quad x_{13} = .31114$

$x_{21} = .75776 \quad x_{22} = .56804 \quad x_{23} = .67420$

Equilibrium Liability Holdings:

$y_{11} = .31846 \quad y_{12} = .51107 \quad y_{13} = .17046$

$y_{21} = .72386 \quad y_{22} = .45497 \quad y_{23} = .82117.$

The algorithm converged in 24 iterations and required 5.09 miliseconds for convergence.

In the fifth and final example, we kept the same tax rate as in Example 4 at $\tau = .3$ but raised the price ceilings to $\hat{r} = 2$, the same level as in Example 1. The numerical results are

Results for Example 5
Equilibrium Prices:

$r_1 = .87731 \quad r_2 = .92179 \quad r_3 = 1.20088$

Equilibrium Asset Holdings:

$x_{11} = .28710 \quad x_{12} = .40066 \quad x_{13} = .31224$

$x_{21} = .75613 \quad x_{22} = .56744 \quad x_{23} = .67643$

Equilibrium Liability Holdings:

$y_{11} = .32066 \quad y_{12} = .51040 \quad y_{13} = .16894$

$y_{21} = .72478 \quad y_{22} = .45746 \quad y_{23} = .81776.$

The algorithm converged in 17 iterations for this example and required 3.59 miliseconds of CPU time for convergence.

In line with (4), for each of the above examples, the algorithm yields asset and liability patterns in which the difference between the total effective volume of an instrument held as an asset is approximately equal to the total volume of the instrument held as a liability when the instrument price is not at one of the bounds. Hence, the market clears for each such instrument, and the price of each instrument is positive in equilibrium.

5. SUMMARY AND CONCLUSIONS

This paper introduces a competitive multi-sector, multi-instrument financial model that explicitly allows for the incorporation of policy interventions in the form of taxes and price controls. The equilibrium conditions for this model are derived using the assumption that the behavior of each sector is one of portfolio optimization. It is then shown that the governing equilibrium conditions are equivalent to a variational inequality problem defined over a compact set. This variational inequality is used to demonstrate existence of the equilibrium asset, liability, and price pattern and to prove that the equilibrium asset and liability holdings are unique.

A decomposition procedure is next proposed for the computation of the equilibrium pattern and convergence established. A notable feature of the algorithm is that it resolves large-scale problems into subproblems of special network structure, each of which can then be solved exactly and in closed form. Finally, several examples are presented illustrating the numerical performance of the algorithm under a variety of policy scenarios. The algorithm required only miliseconds of CPU time for convergence for the particular examples and no more than 25 iterations at a fairly precise level of accuracy. These results suggest that the algorithm should perform well on empirical models, which is a subject of ongoing research.

ACKNOWLEDGEMENTS

The authors are grateful to D.A. Belsley for suggestions that improved the presentation of this work.

This research was funded, in part, by cooperative agreement No. 58–3AEN–0–8066 from the Economic Research Service of the United States Department of Agriculture and by a Faculty Award for Women from the National Science Foundation, NSF Grant No. DMS 9024071.

This research was conducted on the Cornell National Supercomputer Facility, a resource of the Center for Theory and Simulation in Science and Engineering at Cornell University, which is funded in part by the National Science Foundation, New York State, and IBM Corporation.

REFERENCES

Border, K. C. (1985) "Fixed point theorems with applications to economics and game theory", Cambridge University Press, Cambridge, United Kingdom.

Cohen, J. (1987) "The flow of funds in theory and practice, Financial and monetary studies 15", Kluwer Academic Publishers, Dordrecht, the Netherlands.

Dafermos, S. (1990) "Exchange price equilibria and variational inequalities", *Mathematical Programming* **46**, 391–402.

Dafermos, S. and Nagurney, A. (1984) "Sensitivity analysis for the general spatial economics equilibrium problem", *Operations Research* **32**, 1069–1086.

Dafermos, S. and Nagurney, A. (1987) "Oligopolistic and competitive behavior of spatially separated markets", *Regional Science and Urban Economics* **17**, 245–254.

Eydeland, A. and Nagurney, A. (1989) "Progressive equilibration algorithms: the case of linear transaction costs", *Computer Science in Economics and Management* **2**, 197–219.

Florian, M. and Los, M. (1982) "A new look at static spatial price equilibrium problems", *Regional Science and Urban Economics* **12**, 579–597.

Francis, J. C. and Archer, S. H. (1979) "Portfolio analysis", Prentice-Hall, Inc., Englewood Cliffs, New Jersey.

Gabay, D. and Moulin, H. (1980) "On the uniqueness and stability of Nash equilibria in non-cooperative games", in: A. Bensoussan, P. Kleindorfer, and C. S. Tapiero, eds., "Applied stochastic control of econometrics and management science", North-Holland, Amsterdam, The Netherlands.

Hughes, M. and Nagurney, A. (1992) "A network model and algorithm for the analysis and estimation of financial flow of funds", *Computer Science in Economics and Management* **5** 23–39.

Kinderlehrer, D. and Stampacchia, G. (1980) "An introduction to variational inequalities and their applications", Academic Press, New York.

Korpelevich, G. M. (1977) "The extragradient method for finding saddle point and other problems", *Ekonomicheskie i Mathematicheskie Metody* (translated as *Matekon*) **13**, 35–49.

Markowitz, H. M. (1959) "Portfolio selection: efficient diversification of investments", John Wiley and Sons, Inc., New York.

Nagurney, A., Dong, J., and Hughes, M., (1992) "The formulation and computation of general financial equilibrium", *Optimization*, **26**, 339–354.

Nagurney, A. and Zhao, L. (1991) "A network equilibrium formulation of market disequilibrium and variational inequalities", *Networks* **21**, 109–132.

Sharpe, W. (1970) "Portfolio theory and capital markets", McGraw-Hill Book Company, New York.

Thore, S. (1986) "Spatial disequilibrium", *Journal of Regional Science* **26**, 660–675.

Zhao, L. and Dafermos, S. (1991) "General economic equilibrium and variational inequalities", *Operations Research Letters* **10**, 369–376.

AGAPI SOMWARU, V. ELDON BALL AND UTPAL VASAVADA

Modeling Dynamic Resource Adjustment Using Iterative Least Squares

ABSTRACT. The flexible accelerator specification of the demand for capital and labor is estimated using data generated by the United States agriculture. It is assumed that firms maximize the discounted value of expected profits subject to a technology that implies capital and labor stocks are costly to adjust. Given multiproduction behavior, an investment path for the quasi-fixed stocks is developed assuming maximization of the discounted sum of future profits over an infinite horizon.The consistency of data with the adjustment-cost specification requires that the firm's value function be convex in prices. We impose convexity based on the Cholesky factorization of a matrix of constant parameters association with price effects. These matrixes are fitted subject to the condition that they must be non-negative definite.Use non-linear constrained optimization approach for estimating the model the system of quasi-fixed input, variable input, and output equations jointly by inequality constrained iterative least squares method.

1. INTRODUCTION

Econometric studies of producer behavior that exploit the duality between production and cost or profit functions are numerous in the empirical literature. The simultaneous development of duality theory and accessible computational algorithms for nonlinear systems estimation have contributed to the growth of this field of inquiry. Of particular interest to the present study are multiproduct-multifactor models of a production system. Early contributions to the literature include studies by Shumway (1983) and Weaver (1983) which focus on the profit maximizing agricultural firm. An important drawback to these studies is the adoption of a static framework that fails explicitly to model quasi-fixed input adjustment.

In contrast, a recent study by Vasavada and Chambers (1986) proposed an empirical framework to model optimal adjustment of quasi-fixed factors based on well known results in the adjustment-cost literature. The adjustment cost model of a firm is used to rationalize the flexible accelerator specification. Their study adopted the simplifying assumption that the technology was separable in outputs, thereby permitting construction of a single output aggregate. Vasavada and Ball (1988) extends this work to include multiple outputs. Although this study relaxes the separability assumption, the estimated investment demand and output supply equations fail to satisfy the necessary integrability conditions.

This paper improves on previous efforts by incorporating restrictions from theory as part of the maintained model. We illustrate the imposition of curvature and mono-

D. A. Belsley (ed.), Computational Techniques for Econometrics and Economic Analysis, 207–218.

tonicity restrictions on the parameter estimates, not easily handled in conventional estimation approaches. A flexible accelerator specification of the demands for capital and labor is estimated using data from the United States agriculture. It is assumed that firms maximize the discounted value of expected profits subject to a technology that implies capital and labor stocks are costly to adjust. The consistency of data with the adjustment-cost specification requires that the representative firm's value function be convex in prices. We impose convexity based on the matrix of second-order price effects. This matrix is fitted subject to the condition that it should be non-negative definite. We develop and implement a computational procedure for estimating this dynamic production system subject to inequality restrictions implied by convexity in prices.

The paper is organized as follows: Section 2 reviews the relevant theory; the common assumption of static price expectations is adopted. Section 3 describes the functional form specifications. Empirical implementation and data are discussed in Section 4, and the estimation results are discussed in Section 5. Concluding comments are provided in the final section.

2. MODEL OF THE MULTIPRODUCT FIRM

A firm's technology is described by a multiple output production function $Y_{m+1} = F(Y, X, K, \dot{K})$ giving the maximum amount of output Y_{m+1} that can be produced from perfectly variable inputs $X \in R^k_+$ and quasi-fixed stocks $K \in R^n_+$, given that other outputs $Y \in R^m_+$ are produced. The inclusion of $\dot{K}$ as an argument in F indicates the presence of adjustment costs. The production function satisfies $F > 0$; F is twice continuously differentiable; $F_X, F_K > 0$ and $F_{\dot{K}} < 0$; and F is strictly concave in all its arguments. These assumptions are discussed in detail in Epstein (1981).

The firm is assumed to choose an investment path for the quasi-fixed stocks that maximizes the discounted sum of future profits over an infinite horizon:

$$\max_{I \geq 0} \int e^{rt} P^T Y - W^T X - q^T K + F(Y, X, K, \dot{K})\, dt\,, \tag{1}$$

subject to

$$\dot{K} = I - \delta K\,,$$

$$K(0) = K_0 > 0\,,$$

where δ is an $n \times n$ diagonal matrix of positive depreciation rates. $P \in R^m_+$ is the price vector corresponding to Y; $W \in R^k_+$ and $q \in R^n_+$ are rental prices corresponding to X and K, respectively. All prices are measured relative to the price of output Y_{m+1}. Current relative prices are expected to persist indefinitely. A firm may form expectations rationally in this manner when information is costly (Chambers and Lopez, 1984). $r > 0$ is the constant discount rate and r is an appropriately dimensioned scalar matrix with r as the diagonal element. K_0 is the initial endowment of the quasi-fixed factors. Given K_0, the producer chooses a time path $K(t)$, $Y(t)$, $X(t)$, and $Y_{m+1}(t)$, to maximize the present value of rents over an infinite horizon.

Let $J(P, W, q, K)$ denote the optimal value of (1). The Hamilton-Jacobi equation (Arrow and Kurz, 1970) then gives

$$rJ = \max_{I \geq 0} \left[P^T Y - W^T X - q^T K + F(Y, X, K, \dot{K}) + J_K(I - \delta K)\right], \tag{2}$$

where $J_K(\cdot)$ is the vector of shadow prices associated with the quasi-fixed stocks. Under the regularity conditions stated above, Epstein (1981) has shown that the value function J is dual to F and obeys $J_K > 0$; J and J_K are twice continuously differentiable; $(\underline{r} + \delta) J_K + q - J_{KK}\dot{K}^* > 0$; J is convex in (P, W, q); and $rJ - J_K\dot{K}^*$ is convex in (P, W, q). The result that $(\underline{r} + \delta) J_K + q - J_{KK}\dot{K}^* > 0$ restates the equation of motion for J_K implied by the maximum principle and follows by applying the envelope theorem to (2) using the assumption that $F_K > 0$. The statement that $J_K < 0$ follows from the first-order conditions for (2), which imply that $F_{\dot{K}} = -J_K$ and hence the result. Convexity of J in (P, W, q) is intuitively seen by noting that the objective function (1) is the limit of the sum of linear functions in (P, W, q). The requirement that $rJ - J_K\dot{K}^*$ be convex in (P, W, q) is an integrability relationship between J and F. For later use, note that this condition simplifies to convexity of J when J_K is linear in (P, W, q).

The advantage of representing the restrictions implied by dynamic theory in terms of J is its analytical tractability, since the duality between rJ and F implies that the technology can be recaptured by solving

$$\begin{aligned} F^*(Y, X, K, \dot{K}) = \min_{P,W,q} \, & [rJ(P, W, q, K) - P^T Y + \\ & W^T X + q^T K - J_K\dot{K}^*] \, . \end{aligned} \tag{3}$$

When the model generating the data can be approximated by (1), a parametric characterization of optimal policy rules is available. Optimal output supply and investment demand equations are obtained by applying the envelope theorem to (2). Differentiating with respect to (P, W, q) yields

$$Y^* + \underline{r}J_P - J_{PK}\dot{K}^* , \tag{4}$$

$$X^* = \underline{r}J_W + J_{WK}\dot{K}^* , \tag{5}$$

$$\dot{K}^* = J_{qK}^{-1}(\underline{r}J_q + K) \, . \tag{6}$$

The numeraire output obtained from (3) is

$$\begin{aligned} Y_{m+1} = {} & r(J - P^T J_p - W^T J_W - q^T J_q) - \\ & (J_K - P^T J_{PK} - W^T J_{WK} - q^T J_{qK})\, \dot{K}^* \, . \end{aligned} \tag{7}$$

Equations (4) – (7) provide a system of optimal supply and demand equations. Given a characterization of J that satisfies the regularity conditions, these equations provide a straightforward procedure for modelling quasi-fixed input adjustment.

3. MODEL ESTIMATION

To implement the algorithm supplied by equations (4) – (7), a parametric value function must be specified. Consider the following candidate for the value function:

$$J(P,W,q,K) = a_0 + [a_1^T a_2^T a_3^T a_4^T] \begin{bmatrix} P \\ W \\ q \\ K \end{bmatrix}$$

$$+\tfrac{1}{2}[P^T W^T q^T K^T] \begin{vmatrix} A_{11} & A_{12} & A_{13} & A_{14} \\ A_{21} & A_{22} & A_{23} & A_{24} \\ A_{31} & A_{32} & A_{33} & A_{34} \\ A_{41} & A_{42} & A_{43} & A_{44} \end{vmatrix} \begin{bmatrix} P \\ W \\ q \\ K \end{bmatrix} . \tag{8}$$

This is a second-order Taylor series expansion of J in (P, W, q, K). The corresponding output supply and investment demand equations from applying (4) – (7) to (8) are

$$Y^* = \underline{r}[a_1 + A_{11}P + A_{12}W + A_{13}q + A_{14}K] - A_{14}\dot{K} \,, \tag{9}$$

$$X^* = -\underline{r}[a_2 + A_{21}P + A_{22}W + A_{23}q + A_{14}K] + A_{24}\dot{K} \,, \tag{10}$$

$$K^* = (\underline{r} + A_{34}^{-1})\,K + \underline{r}A_{34}^{-1}[a_3 + A_{31}P + A_{32}W + A_{33}q] \,, \tag{11}$$

$$\begin{aligned} Y_{m+1}^* = r[a_0 &+ a_4^T K - \tfrac{1}{2}P^T A_{11}P - \tfrac{1}{2}W^T A_{22}W \\ &-\tfrac{1}{2}q^T A_{33}q + \tfrac{1}{2}K^T A_{44}K - P^T A_{12}W \\ &-P^T A_{13}q - W^T A_{23}q - [a_4^T + K^T A_{44}]\,\dot{K}^* \,. \end{aligned} \tag{12}$$

Before discussing the regularity conditions notice that the net investment equation (11) is a multivariate flexible accelerator (Nadiri and Rosen, 1969) with adjustment matrix $M = (\underline{r} + A_{34}^{-1})$. This can be seen by rewriting (11) as

$$\dot{K}^* = M[\bar{K} - K(P,W,q)] \,, \tag{13}$$

where

$$\bar{K} = -[1 + \underline{r}A_{34}^{-1}]\,\underline{r}[a_3 + A_{31}P + A_{32}W + A_{33}q] \tag{14}$$

is the vector of steady state stocks.

The regularity conditions are readily translated into restrictions on the parameters of J. Since J_K is linear in normalized prices, this curvature condition is equivalent to convexity of J in (P, W, q). This, in turn, is equivalent to non-negative definiteness of the matrix of constant parameters associated with second-order price effects. To impose convexity on the normalized quadratic value function the matrix of constant parameters can be represented in terms of its Cholesky factorization:

$$\begin{vmatrix} A_{11} & A_{12} & A_{13} & A_{14} \\ A_{21} & A_{22} & A_{23} & A_{24} \\ A_{31} & A_{32} & A_{33} & A_{34} \\ A_{41} & A_{42} & A_{43} & A_{44} \end{vmatrix} = LDL^T$$

where L is a unit triangular matrix ($L_{ii} = 1$, $L_{ij} = 0, j > i$) and D is a diagonal matrix. The matrix of parameters is non-negative definite if and only if the diagonal elements (D_{ii}) of the matrix D are non-negative (Rao, 1973). Stability of the optimal adjustment path requires that $(\underline{r} + A_{34}^{-1})$ be a stable matrix, i.e., that all its eigenvalues have negative real parts. It can be shown (Mortensen, 1973) that the net investment equations exhibit properties similar to static factor demands if, in addition, the matrix $A_{34}^{-1}A_{33}$ is symmetric. Moreover, symmetry is sufficient to rule out optimality of cyclical adjustment of the quasi-fixed stocks, which provides further motivation for a test of this specification in the empirical section of the paper.

Two modifications are in order to render the system of equations (9) – (12) estimable. First, a discrete approximation ($\dot{K} \approx K_t - K_{t-1}$) to (11) is used. Second, additive disturbances are appended to reflect a random optimization error. The system of nonlinear equations can then be written:

$$f_{it}(Z_{it}, \Theta) = u_{it}\ , \quad i = m + k + n,\ t = 1, \ldots, T\ , \tag{15}$$

where Z_{it} is a matrix of observed data, Θ is a vector of coefficients to be estimated, and u_{it} is an error of optimization. Assuming that the errors are temporally independent, each with mean zero, identical distributions, and a positive definite variance-covariance matrix Ω, the Aitken-type estimator Θ is obtained by minimizing with respect to Θ

$$S(\Theta) = 1/T \sum_i \sum_t f_{it}(Z_{it}, \Theta)]^T\ (\Omega^{-1} \otimes \underline{1})\ f_{it}(Z_{it}, \Theta)\,. \tag{16}$$

This procedure requires a first-stage estimate of Θ using multivariate least squares, followed by the estimation of Ω based on the inner product of residuals. An estimate of the variance-covariance matrix is then substituted into (16), and $\hat{\Theta}$ is estimated using the new weighting matrix. This procedure is repeated until the parameter vector $\hat{\Theta}$ and the estimated variance-covariance matrix $\hat{\Omega}$ stabilize. For the Aitken estimator, this iterative procedure yields an asymptotically efficient estimator, which is otherwise not the case (Malinvaud, 1970).

The estimation procedure used in this study must be capable of imposing the theoretical curvature restrictions on the value function. This procedure is characterized by the non-linear constrained optimization problem

$$S(\Theta) = \min_{\Theta} 1/T \sum_{i} \sum_{t} f_{it}(Z_{it}, \Theta)^T \, (\Omega^{-1} \otimes \underline{1}) \, f_{it}(Z_{it}, \Theta) \, . \tag{17}$$

subject to

$$h_i(\Theta) \geq 0 \, , \quad i = 1, \ldots, m + k + n \, ,$$

where $h_i(\Theta)$ are non-negativity restrictions based on a Cholesky factorization of a real symmetric matrix. The objective function (17) is minimized subject to the constraint set. The model specification and estimation are accomplished using the General Algebraic Modeling System (GAMS version 2.25) (Brooke, Kendrick and Meeraus, 1988) on the Theory Center's Supercomputer (CNSF) at Cornell University. Numerical solutions to this problem are computer intensive and the CNSF was an extremely valuable resource for this purpose.

Using GAMS, we modify the Aitken procedure by imposing inequality constraints in the first stage (Ruble, 1968), replacing $(\hat{\Omega}^{-1} \otimes \underline{1})$ in (17) with the identity matrix and then solving for $\hat{\Theta}$. Using $\hat{\Theta}$, a new estimate of $\hat{\Omega}$ is obtained and the problem (17) is solved with $(\hat{\Omega}^{-1} \otimes \underline{1})$ employed as the weighting matrix. Finally, we iterate over steps one and two until the estimated parameter vector $\hat{\Theta}$ and error variance-convariance matrix $\hat{\Omega}$ stabilize (Iterative Least Squares).

4. EMPIRICAL IMPLEMENTATION AND DATA

The empirical model identifies two output categories including livestock and crops. The stocks of durable equipment, service buildings, land, farm-produced durables, and self-employed labor are assumed costly to adjust. Hired labor, energy, and other purchased inputs are assumed to adjust freely to current prices and stocks of the quasi-fixed inputs.

The output series are defined as the quantities marketed (including unredeemed Commodity Credit Corporation loans) plus changes in farmer-owned inventories and quantities consumed by farm household. The indexes of output are based on value to the producer. For this reason, commodity prices are adjusted to reflect direct payments to producers under government programs.

The labor data were developed by Gollop and Jorgenson (1980). They disaggregate labor input and labor cost into cells cross-classified by two sexes, eight age groups, five education groups, two employment classes (hired and self-employed), and ten occupational groups.

The value of labor services equals the value of labor payments plus the imputed value of self-employed and unpaid family labor. The imputed wage rate is set equal to the mean wage rate of hired farm workers with the same occupational and demographic characteristics.

The capital input data are derived from information on investment and the outlay on capital services. There are twelve investment series used to calculate capital stocks. The perpetual inventory method (Jorgenson, 1974) is used, and the service lives are those of Bulletin F published by the U. S. Treasury Department. Rental prices for each asset are constructed taking account of variations in effective tax rates and rates of return, depreciation, and capital gains. The value of capital services is computed as the product of the rental price and the quantity of capital at the end of the previous period. A more detailed discussion of the procedures used in constructing the capital price and quantity series is found in Ball (1985).

5. ESTIMATION RESULTS

The system of quasi-fixed input, variable input, and output equations is jointly estimated by an inequality-constrained iterative least squares method, which is equivalent to the maximum likelihood method. Parametric inequality constraints associated with the convexity of J in (P, W, q) are imposed during estimation. When these restrictions are true, the resulting estimates are asymptotically efficient. Further justification for imposing these constraints comes from noting that the derived structural model is itself an implication of economic theory. It would, therefore, be inconsistent selectively to utilize only the structural model implied from theory while rejecting associated parametric restrictions.

This highly nonlinear model has 788 variables, 697 equations and 8582 non zero elements. 4.71 Mbytes are required for execution of each iteration on the variance-covariance matrix.

Adjustment Matrix

Point estimates of adjustment parameters are reported in Table 1. Since the accepted model did not preclude interdependent adjustments, off-diagonal elements of this important matrix are non-zero. A positive off-diagonal element, say for M_{13}, indicates that when input 3 is below its long-run value, disinvestment in input 1 is induced. Similarly, a negative value for the same elements of the adjustment matrix implies that, under identical circumstances, investment in input 1 will be induced. In this fashion, numerical values of off-diagonal elements reflect structure of interdependent adjustments in aggregate U.S. agriculture.

Now turn to the diagonal elements of the adjustment matrix. Consider the element M_{11}. A value of –0.160 for this coefficient suggests that, when actual stocks for durable equipment diverge from long-run values, it takes a little over six years to complete the needed adjustment, given that all other inputs are at long-run equilibrium levels. Similarly, M_{22}, M_{33}, and M_{44} supply an interpretation of adjustment speeds for other quasi-fixed inputs.

Real estate takes approximately fifteen years to adjust, while farm-produced durables take only two years. By contrast, the adjustment speed for self-employed and unpaid family labor is close to nine years to adjust. Values for remaining inputs are, for the most part, numerically similar. One important conclusion emerging from the multiple-input, multiple-output model is that disaggregation of labor into two categories provides more reasonable results. Earlier investment studies derived high adjustment lags for labor when complete supply response systems are estimated (Vasavada and Chambers, 1986). Since self-employed labor and hired labor have qualitatively different characteristics, such a disaggregation scheme improves

TABLE 1
Estimated Adjustment Matrix.

Parameter	Estimate	Parameter	Estimate
$M(1,1)$	–0.160	$M(3,1)$	0.824
$M(1,2)$	–0.127	$M(3,2)$	–0.050
$M(1,3)$	0.182	$M(3,3)$	–0.554
$M(1,4)$	–0.060	$M(3,4)$	–0.326
$M(2,1)$	–0.035	$M(4,1)$	0.045
$M(2,2)$	–0.060	$M(4,2)$	–0.161
$M(2,3)$	0.019	$M(4,3)$	0.073
$M(2,4)$	0.036	$M(4,4)$	–0.134

Note: 1 is durable equipment and service buildings, 2 is real estate, 3 is farm produced durables, and 4 is self-employed and unpaid family labor.

specification in applied dynamic production models (Gunter and Vasavada, 1988).

Elasticities: Short-Run
Point estimates for the model reported in Table 1 can be used to evaluate short-run elasticities. One important feature of the adjustment cost model is its emphasis on maintaining a clear conceptual distinction between short-run and long-run responses to changing opportunity costs. This was cited earlier as an important justification for conducting the present study. Since agricultural producers are exposed to constantly changing opportunity costs, it is useful to evaluate behavior in this volatile economic environment.

Short-run elasticities are reported in Table 2. Diagonal elements of the matrix are own-price elasticities; off-diagonal elements are the corresponding cross-price elasticities. Own-price elasticities for the quasi-fixed inputs are observed to be numerically small in magnitude. This suggests that, given the technological constraints faced by agricultural producers, a change in the rental price does not evoke significantly different utilization patterns for these inputs in the short-run. A different conclusion emerges when short-run elasticities of variable inputs are evaluated. A change in the price of variable inputs, namely hired labor and purchased inputs, induces shifts in utilization patterns. Now turn to an examination of own-price elasticities for outputs. This value for livestock is extremely small although values for all other inputs exceed one. Among outputs considered, the own-price elasticity for grain was highest. Values for dairy and other crops are numerically similar.

6. CONCLUSIONS AND POLICY IMPLICATIONS

It is possible to distinguish several approaches to econometric model-building. First, the economist can use *ad hoc* restrictions to specify a structural model. As an

TABLE 2
Short-run elasticities with respect to prices.

Stocks	Prices								
	Equipment & Buildings	Real Estate	Farm Produced Durables	Self-Employed Labor	Hired Labor	Energy	Purchased Inputs	Livestock	Crops
Equip. % Bldg.	–1.289E–6	–3.265E–7	4.486E–6	–1.039E–6	–2.957E–6	–7.063E–7	–1.119E–6	–3.456E–6	6.409E–6
Real Estate	–7.848E–8	–1.359E–8	2.184E–7	9.635E–8	1.827E–7	–2.799E–8	–1.859E–8	–4.208E–7	6.202E–8
Farm Durables	5.903E–8	–2.205E–7	–8.3221E–6	–3.5105E–6	–3.410E–6	–4.209E–8	–5.045E–7	3.927E–6	1.202E–5
Self Employed Labor	–9.0371E–7	–3.5204E–7	5.0414E–7	–2.319E–6	–3.957E–6	–5.293E–7	–1.0848E–6	–2.961E–7	8.938E–6
Hired Labor	–0.020	–0.005	–0.012	–0.062	–0.131	–0.016	–0.021	–0.039	0.277
Energy	–0.022	–0.005	–0.013	–0.025	–0.039	–0.014	–0.017	–0.093	0.241
Purchased Inputs	–0.005	–0.002	–0.007	–0.011	–0.015	–0.003	–0.005	–0.008	0.130
Livestock	0.016	0.002	0.011	0.006	0.004	0.009	0.008	0.105	–0.146
Crops	0.283	0.0198	0.626	0.277	0.397	0.175	0.134	0.727	0.001

alternative to building *ad hoc* structural models, one can model dynamic relationships between economic variables as vector autoregressions. These are essentially reduced-form relationships, which utilize *a priori* restrictions only parsimoniously. The model utilized in the present study is a structural one, with a structure derived *explicitly* from relevant economic theory. Restrictions placed on the model are not *ad hoc*, but rather, are shown to be implied by the value maximization hypotheses. In this sense, there is significant justification for adopting this approach. However, there is always the risk that restrictions imposed on the model, especially curvature, are not supported by the data. It is argued here that structural models based on dynamic optimization should be used in their entirety or not at all. Selective utilization of econometric equations from value maximization without implied parametric restrictions is hard to justify. This is equally true for static models within the dual framework, where curvature conditions are frequently not imposed. An important exception is Ball (1988).

Accordingly, a methodology is developed to impose convexity restrictions on a multiple-input, multiple-output model of aggregate agricultural production. Point estimates of parameters are used to evaluate short-run elasticities. Generally, short-run behavioral responses of quasi-fixed inputs to opportunity costs are small. Hence, implementation of a favorable tax policy in U.S. agriculture would have a small, albeit non-negligible, effect on investment. By contrast, changing relative prices are observed to have a significant effect on quantities of variable inputs employed. In a similar fashion, supply response is also observed to be elastic. A supply management policy based on manipulating market incentives can prove to be effective, especially in U.S. agriculture.

Several shortcomings of the present modeling effort may be mentioned. First, it is necessary to address the issue of expectations formation in a more sophisticated manner than was possible with the present model. Inclusion of non-static expectations would serve to strengthen the foundations of the multiple-input, multiple-output model. This will necessarily entail imposition of highly complex cross-equation restrictions. For this reason, incorporating non-static expectations is relegated to a future goal. A second issue that needs to be addressed is the inclusion of U.S. agricultural policy variables in a more direct fashion than was possible within the exiting framework. While recognizing the importance of this objective, it should be noted that the U.S. agricultural production sector has been exposed to a diverse set of policy instruments, which have changed constantly over time. It is by no means an easy task to integrate this matrix of policy interventions into the existing model. Future investigations must, therefore, concentrate on the dual objective of both improving model specification and making models more applicable to policy analysts and performing hypotheses tests on the structure of the production system. Finally, alternative functional forms must be subject to experiment to evaluate robustness of empirical results to alternative specifications (Baffes and Vasavada, 1989). These are some meaningful directions to pursue in the study of resource response to changing market incentives.

REFERENCES

1. Arrow, K. and M. Kurz (1970), *Public Investment, the Rate of Return and Optimal Social Policy*, Baltimore; Johns Hopkins Press.
2. Bacharach, M. (1965), "Estimating Non-Negative Matrices From Marginal Data," *International Economic Review*, 6: 294–310.
3. Baffes, J. and U. Vasavada (1989), "On the Choice of Functional Forms in Agricultural Production Analysis," *Applied Economics*, 21: 1053–1061.
4. Ball, V. E. (1988), "Modelling Supply Response in a Multiproduct Framework," *American Journal of Agriculture Economics*, 70: 813–25.
5. Ball, V.E. (1985), "Output, Input, and Productivity Measurement in U.S. Agriculture," *American Journal of Agricultural Economics*, 67: 475–86.
6. Brooke, A. D. Kendrick, and A. Meeraus (1988), *GAMS: A User's Guide*, The Scientic Press.
7. Epstein, L. G. (1981), "Duality Theory and Functional Forms for Dynamic Factor Demands," *Review of Economic Studies*, 48: 81–95.
8. Chambers, R. G. and R. Lopez (1984), "A General Dynamic Supply Response Model," *Northeastern Journal of Agricultural and Resource Economics*, 13: 142–154.
9. Gollop, F., and D. Jorgenson (1980), "U.S. Productivity Growth by Industry, 1947–73." in *New Developments in Productivity Analysis*, ed. J. Kendrick and B. Vaccara. National Bureau of Economic Research, Studies in Income and Wealth, Vol. 44. Chicago; University of Chicago Press.
10. Gunter, L. and U. Vasavada (1988), "Dynamic Labor Demand Schedules for U.S. Agriculture," *Applied Economics.*
11. Jorgenson, Dale W. (1974), "The Economic Theory of Replacement and Depreciation," *Econometrics and Economic Theory: Essays in Honor of Jan Tinbergen*. Ed. W. Sellakaerts; London, MacMillian Publishing Co., pp. 189–222.
12. Lau, L. J. (1978), "Testing and Imposing Monotonicity, Convexity, and Quais-Convexity Constraints," *Production Economics: A Dual Approach to Theory and Applications*. Ed. D. McFadden and M. Fuss; Amsterdam, North Holland.
13. Malinvaud, E. (1970), *Statistical Methods of Econometrics*, Amsterdam, North Holland, 1970.
14. Mortenson, D.T. (1973), "Generalized Cost of Adjustment and Dynamic Factor Demand Theory," *Econometrica*, 41: 657:65.
15. Murtaugh B. and M. Saunders (1983), *MINOS 5.0 Users Guide*. Systems Optimization Laboratory Technical Report SOL 83–20, Stanford University, Stanford, Ca.
16. Murtaugh B. and M. Saunders (1983), *MINOS 5.0 Users Guide*. Systems Optimization Laboratory Technical Report SOL 83–20, Stanford University, Stanford, Ca.
17. Nadiri, M. I., and S. Rosen (1969), "Interrelated Factor Demands," *American Economic Review*, 59: 457–71.
18. Rao, C.R. (1973), *Linear Statistical Inference and Its Applications*, John Wiley and Sons; New York.
19. Ruble, W.L. (1968), "Improving the Computation of Simultaneous Stochastic Linear Equation Estimates," Ph.D. Thesis, Department of Economics Michigan State University; East Lansing.
20. Shumway, G.R. (1983), "Supply, Demand, and Technology in a Multiproduct Industry: Texas Field Crops," *American Journal of Agricultural Economics*, 65: 748–60.
21. U.S. Treasury Department (1942), Bureau of Internal Revenue, Income Tax Depreciation and Obsolescence, Estimated Useful Life and Depreciation Rates, Bulletin F; Washington, DC.
22. Vasavada, U., and V.E. Ball (1988), "Modeling Dynamic Adjustment in a Multi-Output Framework," *Agricultural Economics.*
23. Vasavada, U., and R. G. Chambers (1986), "Investment in U.S. Agriculture," *American Journal of Agricultural Economics*, 68: 950–60.

24. Weaver. R.D. (1983), "Multiple Input, Multiple Output Production Choices and Technology in the U.S. Wheat Regions," *American Journal of Agricultural Economics*, 65: 45–56.

ATREYA CHAKRABORTY AND CHRISTOPHER F. BAUM*

Intensity of Takeover Defenses: The Empirical Evidence

ABSTRACT. This paper focuses on the construction of an index of the intensity of firms' antitakeover defenses. While many aspects of corporate behavior are qualitative in nature, an evaluation of a firm's stance and the underlying motives for its behavior often depend on the elements of a set of qualitative factors. The interactions beween these factors are likely to have important implications. In this context, only a composite measure will capture these interactions and their implications for firms' actions. We focus on the creation of an ordinal measure of anti-takeover defenses and utilize the ordered probit estimation technique to relate the magnitude of this measure to the motives for instituting these defenses. Our estimates are generally supportive of the managerial entrenchment hypothesis.

1. INTRODUCTION

The evaluation of corporate behavior generally involves the assessment of a firm's stance toward its competitors in the markets in which it operates. Whether we are examining the firm's behavior in the product markets, the factor markets or the market for corporate control, we would expect that measurement of the firm's stance involves both quantitative and qualitative aspects. Much of financial research has focused upon the readily quantifiable aspects of behavior. However, if we focus on the market for corporate control, it is apparent that many aspects of a firm's stance can only be reflected in qualitative factors: for instance, does the firm offer "golden parachute" severance contracts to its executives? Since the presence or absence of such a clause in an executive's contract is likely to have important incentive effects on their behavior, it is the qualitative aspect that we must study.

While techniques such as binomial logit or probit may be readily applied to individual qualitative measures and their causal factors, such models fall short of capturing the essence of many legal aspects of a firm's stance. For instance, the recent wave of "anti-takeover amendments" adopted by major American corporations has left us with an entire array of qualitative factors at the firm level. Firm ABC may have adopted takover defenses 1, 4, and 6, whereas its rival Firm XYZ may have chosen to adopt defenses 2, 6, and 9. Which firm has the stronger defenses against an hostile takeover? As in the evaluation of any deterrent capability, such a question is difficult to answer.

* We thank Robert Taggart, Jr., Richard Arnott, Stephen Polasky, seminar participants at Université du Québec à Montréal, and participants in the 1992 meetings of the Society for Economic Dynamics and Control and the Financial Management Association for their valuable comments, criticisms and many insightful discussions. The usual disclaimer applies.

D. A. Belsley (ed.), Computational Techniques for Econometrics and Economic Analysis, 219–231.

In this study, we attempt to resolve this quandary by constructing an ordinal index of the strength of firms' takeover defenses which should enable us to categorize firms by "intensity." Although this technique is applied specifically to the study of firms' takeover defenses, it should be evident that it could equally well be applied to many sets of qualitative aspects of firms' behavior–for instance, in evaluating their marketing strategy, the various dimensions of their research and development effort, or the importance of various types of intangible assets on their balance sheet.

The plan of the paper is as follows: the next section gives a brief summary of the types of anti-takeover amendments and discusses some of the evidence in the literature regarding their effects on corporate performance. In Section 3, the composite measure of intensity is developed in the context of a dataset of 68 U.S. corporations, and the ordered probit technique is described. Section 4 specifies an hypothesis under which the intensity of anti-takeover defenses should be related to various causal factors and presents our empirical findings from the ordered probit model. Section 5 concludes and presents suggestions for future research.

2. ANTI-TAKEOVER AMENDMENTS IN AMERICAN CORPORATIONS

The merger wave of the 1980s, coupled with the sophistication of investment banks' financial engineers, caused many large corporations in the United States to include anti-takeover amendments in their corporate charters. Rosenbaum (1986) details anti-takeover measures for 424 Fortune 500 firms as of May 1986. Among these companies, a surprisingly large number (403) had at least some amendments that were designed to have anti-takeover consequences or that could be adopted to thwart takeover attempts. Of these firms, Rosenbaum documents 143 as having poison pills, 158 with fair price amendments, 223 with classified boards, 362 with blank check provisions, 65 that require a supermajority to approve a merger, and 222 firms as having some types of limited shareholder rights. We now define each of these defensive measures.

Poison Pills: These are preferred stock rights plans adopted by the management, generally without the shareholders' approval. These amendments are exclusively tailored to thwart hostile bids by triggering actions that make the target financially unattractive.

Fair Price Amendments: These are designed to prevent two-tier takeover offers. They require that the bidders pay all the tendering shareholders the same price. Most fair price provisions can be waived if the bidder's offer is approved by a supermajority of target shareholders. This supermajority requirement may be as low as 66% or as high as 90%.

Classified Boards: Such amendments divide the board of directors into three classes. Each year only one class of directors is due for election. This prevents a raider from immediately replacing the full board and taking control of a company, even if the raider controls a majority of the shares. More importantly, such amendments also make proxy contests over control extremely difficult.

Blank Check: These give the managers (via the board of directors) a "very broad discretion to establish voting, dividend conversion and other rights for preferred stock that a company may use" (Rosenbaum, 1986, p. 7). Such discretionary powers may easily be used to issue securities primarily intended to thwart takeovers (poison pills). Finally, since the SEC requires companies seeking to issue preferred shares to disclose to shareholders that unused preferred stock may have anti-takeover effects, regardless of the company's professed intention, Rosenbaum contends that blank checks should indeed be classified as anti-takeover measures.

Supermajority requirements: These provisions require approval for specified actions that is far higher than that set by state laws. Actions which would normally require majority approval, under these amendments, require approval levels as high as 90%. Such provisions require a hostile bidder to obtain higher percentages of shares to obtain control over a firm.

Dual Class Recapitalization: A new class of equity is distributed to shareholders with superior voting rights but inferior dividends or marketability. This permits incumbent managers to obtain a majority of votes without owning a majority of the common shares.

Limiting Shareholders' Rights: These include provisions such as: no shareholder action by written consent (without a meeting), procedural requirements for shareholders to nominate directors, restricting shareholders from calling special meetings, and supermajority requirements to repeal classified boards. Each of these measures selectively, or in conjunction with other measures, can be used to deter the shareholders from facilitating changes in corporate control.

Studies investigating the effects of anti-takeover amendments have been, on the whole, quite inconclusive about the net effect of these financial innovations. DeAngelo and Rice (1983), while examining a sample of NYSE listed firms adopting anti-takeover amendments during 1971–79, found statistically insignificant (albeit negative) abnormal stock returns around the announcement of such amendments. Conversely, Linn and McConnell's (1983) investigation into abnormal returns at the announcement date for 475 NYSE listed firms (between 1960–80) found significantly positive abnormal returns. Malatesta and Walkling (1988) report statistically significant reductions of shareholders' wealth for firms that adopt poison pill defenses. They also note that firms adopting such defenses were significantly less profitable than the average firm in their industries during the year prior to adoption. Jarrell and Poulson (1987), investigating similar reactions for 600 firms over the period 1979–85, detected a significantly negative price reaction for certain kinds of amendments. Pugh et al. (1992), investigating the impact of charter amendments' adoptions on firms' capital expenditures and R&D outlays, conclude that managers take a longer-term view following such actions. In light of these results, it is difficult unambiguously to determine whether the adoption of ATAs is consistent with shareholders' interests.

3. A MEASURE OF INTENSITY OF ANTI-TAKEOVER AMENDMENTS

Many researchers' analyses of takeover defenses have been hindered by the general unavailability of data on the prevalence of such measures. Although a detailed study of a firm's SEC filings and annual reports would provide much of this information, there are still serious issues of heterogeneity and classification of the firm-specific measures into generally accepted categories.

Our goal is an index that incorporates all categories of defensive strategies into a single ordinal index value. The presence of a particular anti-takeover defense in a corporate charter is a qualitative factor. Since there is no consensus on the severity of various defensive mechanisms, it would appear that any cardinal index of the strength of these defenses would be arbitrary. To deal with this critique, we have used the qualitative information in Rosenbaum's Takeover Defenses dataset, combined with other measures of firms' characteristics, to build an ordinal index of "intensity."

We combine data from Rosenbaum's dataset, which indicates whether firms had various anti-takeover amendments in place as of 1986, with firms' characteristics from Thies and Baum's Panel84 dataset.[1] The latter dataset contains annual data on the firm level for 1977–1983 for a total of 134 large U.S. manufacturing corporations. Panel84 reconstructs financial statements on a replacement-cost-accounting basis, exploiting inflation-adjusted data obtained from firms' Forms 10-K and annual report disclosures required during this period by the Securities and Exchange Commission and the Federal Accounting Standards Board.[2] These data are particularly appropriate for this study since we hypothesize that Tobin's q is a relevant explanatory variable, and Panel84 contains consistent estimates of Tobin's q that are largely free from the imputation bias created by the commonly-used methods of Brainard, Shoven and Weiss (1980). This bias may be especially harmful, as it represents measurement error correlated with common indicators of firm performance, as noted by Klock et al. (1991).

Of the 134 firms in Panel84, there are 68 which are also to be found in Rosenbaum's Takeover Defense dataset.[3] We use these matching firms in the empirical analysis of the next section. Since the Panel84 data provides us with annual detail of firms' performances from 1977 through 1983, it can be viewed as exogenous to the observation of firms' takeover defenses in 1986. Although the intervening years could provide useful information, the use of firm characteristics from the 1977–1983

[1] An earlier version of the dataset, containing 100 firms, is documented in Thies and Sturrock (1987), and was further described in Klock, Thies and Baum (1991). Of these 100 firms, 98 appear in the present dataset. Data for the additional 36 firms were gathered by Glenn Rudebusch and Steven Oliner of the Federal Reserve Board of Governors.

[2] During the period in which firms were required to report current cost data in their annual reports (1976–1983), they were given broad leeway in the methodology used for their calculations. Thus, the accuracy of these figures could be debated. Nevertheless, we presume that these estimates of replacement cost are likely to be more reliable than those which could be constructed by outside researchers via adjustments to historical costs, using aggregate price deflators for capital goods, without access to firm-specific vintage data.

[3] A list of the 68 firms and their two-digit industry codes is available from the authors on request.

TABLE 1
Descriptive statistics of the sample.

Variable	Mean	Std. Dev.	Minimum	Maximum
Tobin's Q	0.97	0.43	0.45	2.83
Financial Leverage	0.43	0.34	0.02	1.50
R&D per dollar of Sales	0.025	0.026	0.0	0.110
Advertising per dollar of Sales	0.015	0.027	0.0	0.166
Sigma from CAPM regression	7.58	2.74	4.45	19.94
CAPM Beta	1.04		0.46	2.49
Sales, $ Bil.	3.330		0.196	62.790
Cash Flow, $ Bil.	0.215		–0.003	4.398
Net Income, $ Bil.	0.086		–0.332	1.087
Total Assets, $ Bil.	3.221		0.185	46.456
Pr{Poison Pill}	0.29	0.46		
Pr{Blank Check}	0.85	0.36		
Pr{Classified Board}	0.62	0.49		
Pr{Fair Price Amendment}	0.41	0.50		
Pr{Supermajority for Merger}	0.19	0.40		
Pr{Dual Class Recapitalization}	0.09	0.29		
Pr{Limited Shareholder Rights}	0.60	0.49		

Notes:
Statistics (other than the ATA probabilities) are based on the firm average values for 1977–1983. There are 68 firms in the analysis.

period allows us to avoid possible simultaneity between firms' adoption of takeover defenses and the financial markets' reactions to those events. Of course, our knowledge that a certain defense is in place in 1986 does not tie down the date of its adoption, but the use of panel data from a seven-year window should mitigate the "announcement effects" if a defense was first put in place during this seven-year period. Descriptive statistics based on the time-series average values of the explanatory variables used in our analysis are presented in Table 1. A number of other variables' descriptive statistics from this seven-year period are given to illustrate the range of firms considered in this sample. We also present the sample proportions of firms who have adopted the various anti-takeover amendments described in the previous section.

The takeover defenses are classed, via extensions to Ruback's (1988) scheme, as either innately mild or severe. The "mild" defenses are those in the set Fair Price, Supermajority, Blank Check, and Limited Rights. The "severe" defenses are, according to Ruback, Poison Pill, Dual Class Recapitalization, and Classified Board. The most disaggregate form of the ordinal index is defined in terms of the number of "mild" defenses (NRMILD) and the number of "severe" defenses (NRSEV) as

Index value	Definition	Number of firms
0	NRMILD = 0 and NRSEV = 0	2
1	NRMILD > 0 and NRSEV = 0	17
2	NRMILD ≤ 3 and NRSEV = 1	28
3	(NRMILD > 3 and NRSEV = 1) or NRSEV ≥ 2.	21

Although this method yields an ordinal measure of the intensity of anti-takeover amendments, the problem still remains. How might this ordinal measure be related to a set of causal factors so that an hypothesis on the intensity of ATAs might be rigorously tested? The solution to this problem is to be found in the econometric technique of ordered probit analysis (Zavoina and McElvey, 1975). Standard binomial probit models relate the probability of observing a particular attribute to a set of causal factors via a nonlinear transformation: the cumulative distribution function of the error process. Multinomial probit models allow for multiple, mutually exclusive and exhaustive alternatives, but make no assumption about the ordinality of those alternatives (e.g. choosing to commute via train is not ambiguously preferred to the bus or the car-pool). Here, however there is more information from the values of the ordinal index, and even though merely ordinal, we can relate these values to an unobservable "latent variable" in an extension of the binomial probit technique.

In binomial probit, the explanatory variables (and a Gaussian error term) are linearly related to an unobservable latent variable, or index, as $I_i = \mathbf{X}\beta + \varepsilon_i$. If that index equals or exceeds a threshold value, we observe the attribute ($y_i = 1$); if it falls short of the threshold, we do not ($y_i = 0$). While the effect of an explanatory variable on the index is linear, the effect on the predicted probability of observing the attribute is not, since $\Pr[y_i = 1] = F(I_i)$, where F is the cumulative distribution function of the normal distribution.

The ordered probit technique follows the same approach, but assumes that there are multiple thresholds that the latent variable may cross. The basic model is:

$$z = \mathbf{X}\beta + \varepsilon\,, \quad \varepsilon \sim N[0,1]\,,$$

where

$$y = 0 \;\text{ if }\; z < \mu_0$$

$$y = 1 \;\text{ if }\; \mu_0 < z < \mu_1$$

$$y = 2 \;\text{ if }\; \mu_1 < z < \mu_2$$

$$\ldots$$

$$y = J \;\text{ if }\; z > \mu_{j-1}\,.$$

Here y is observed, while z is not. We wish to estimate the parameters of the β vector as well as the vector μ. Since the model includes a constant term, one of the μ's is not identified. We accordingly normalize μ_0 to zero, and estimate $\mu_1 \ldots \mu_{J-1}$. In our setting, J is equal to 3, the maximum value of intensity, so that estimates of μ_1 and μ_2 may be recovered. As in the binomial probit model, we cannot separately identify the variance of the error term, which is thus set to unity. The ordered probit estimator is available as a component of LIMDEP (Greene, 1992) and is further described in Greene (1990, pp. 703–706).

4. ATA INTENSITY AND FIRMS' CHARACTERISTICS

In this section, we formulate an hypothesis under which the intensity of firms' anti-takeover amendments should be related to observable measures of their behavior. If we observe firms altering their corporate charters to include these amendments – as we did in large number in the 1980s – we might conclude that firms' actions are merely reflective of their shareholders' interests and that these actions are being taken to maximize the value of the firm, as neoclassical theory would suggest.

An alternative explanation has been provided by finance researchers analyzing the aspects of agency costs. In this formulation, protective measures such as ATAs may well be indications of an "entrenched management" – essentially, managers who are looking after their personal interests first and foremost and may indeed take actions contrary to shareholders' best interests. In examining the intensity of ATAs, we consider explanatory factors that would be indicative of the entrenched management hypothesis. Under this hypothesis, we assume that the primary role of anti-takeover amendments is to insulate underperforming managers from the discipline of the market for corporate control.

The entrenched management hypothesis suggests that firms that are more likely to be takeover targets should display a greater intensity of takeover defenses. What gives rise to this vulnerability? One obvious cause is a poor performance – irrespective of reason – that leads the firm's valuation in the financial markets to be low. Since the financial markets' evaluations of a firm's worth are forward-looking measures, a firm with poor performance and poor prospects will have a low valuation in the stockmarket and may well be a takeover target – especially if a raider considers the root of that poor performance to be inadequate management. The shareholders of such a firm would not want to place obstacles in the raider's path, and if they share the market's low opinion of managerial talent, they would encourage a takeover. Such managers would, naturally, have every reason to protect themselves, most especially in the case where they recognize their own shortcomings.

To quantify this rationale, we utilize Tobin's q as an objective and non-myopic indicator of current management's performance. Past research on the relation between q and takeovers (Servaes, 1991; Lang et al., 1989; Morck et al., 1988) has interpreted q as a measure of managerial performance: e.g. "In general the shareholders of low q targets benefit more from takeovers than shareholders of high q targets." (Lang et al., 1989, p.137).

If q were indeed a good proxy for managerial efficiency, and if defensive strategies were primarily instituted for shareholders' interests, then one should not expect Tobin's q to have any explanatory power in predicting the adoption of defensive measures. However, if the manager were in charge of instituting defensive instruments, one would expect an inverse relationship between anti-takeover amendments and Tobin's q: the less efficient the manager (as signalled by a lower q), the greater would be the need for her insulating herself from the disciplining forces of the market.

The second explanatory variable which we include in the model of ATA intensity is a measure of financial leverage. The rationale for including financial leverage as an explanatory variable for the analysis of takeover defenses comes from the works of Jensen (1986) and Ross (1977). The role of debt in motivating organizational efficiency is well documented in the principal-agent literature. Jensen (1986) points to the possibility that, more than any other action, debt creation can actually lend credibility to the management's promise to pay out future cash flows. Thus, higher levels of debt lower the agency cost of cash flows by reducing the discretionary levels of cash flow available to the management. Conversely, managers less willing to be exposed to the disciplines of the financial markets would tend to have lower levels of debt and greater need for defensive measures to insulate them from the market for corporate control.

The explanatory power of leverage in predicting anti-takeover amendments should be minimal if we assume that these amendments primarily serve the shareholders' interests. We should not expect any systematic relationship between leverage and defensive measures under this hypothesis if we assume a well functioning market for corporate control. However, if defensive amendments serve primarily the interests of an entrenched management, ceteris paribus, one would expect an inverse relationship between leverage and the probability of adopting anti-takeover measures.

Under the hypothesis that the adoption of anti-takeover amendments reflects the actions of an entrenched management, we would expect to find

$$\Pr\{\text{ATA}\} = f\,[\,\overset{[-]}{\text{Tobin's q}},\ \overset{[-]}{\text{Leverage}}\,]$$

with respect to the likelihood of observing individual ATAs, or, by making use of the index of ATA intensity, that

$$\text{Intensity}\{\text{ATA}\} = f\,[\,\overset{[-]}{\text{Tobin's q}},\ \overset{[-]}{\text{Leverage}}\,].$$

The explanatory variables are defined as a market-value-based Tobin's q measure, Q–AVG, which is the simple average of Tobin's q from the Panel84 dataset for the years 1977–1983, as well as LEV–AVG, the firm's average financial leverage (debt/equity ratio)[4] over the seven years.

The first column of Table 2 presents the ordered probit estimates of Intensity. The model is quite successful, with both Tobin's q and financial leverage possessing significantly negative coefficient estimates, as predicted by the managerial entrenchment hypothesis.

---

[4] Measured in market (or fair) value terms.

TABLE 2
Ordered probit estimates of the intensity of takeover defense strategies.

Dependent variable	Intensity	Intensity2
Constant	5.396	4.997
	(2.7)	(2.6)
Q–AVG	–2.110	–1.897
	(–3.5)	(–3.1)
LEV–AVG	–1.351	–1.201
	(–2.6)	(–2.2)
μ_1	2.134	2.026
	(1.3)	(1.4)
μ_2	3.395	
	(2.0)	
Log-likelihood	–67.666	–37.939
Model χ^2	24.947 (0.5×10^{-7})[a]	17.474 (0.00001)[a]

Notes:
Estimates are based on 68 observations. Asymptotic t-statistics are given in parentheses below the estimated coefficients. The model χ^2-statistic tests the hypothesis that all slopes are zero; its tail probability (p-value) is given in parentheses. "a" denotes significance at the one per cent level.

To ensure that the validity of these results is not an artifact of the Intensity index, we constructed an aggregation of the index, collapsing categories 2 and 3 into a single category. The resulting Intensity2 index considers that all firms with one or more "severe" anti-takeover amendments are given the same value. The results from this version of the ordinal index are given in the second column of Table 2. They are qualitatively very similar to those achieved with Intensity: Tobin's q and leverage are highly significant with the expected negative sign.

In practice, the value of such a model of Intensity must involve its predictive power. Table 3 reports the distribution of "actual" values of the original Intensity index, and presents the distribution of predicted values for the model given in Table 2. The discrete prediction of the model, for each firm, is taken as the alternative among the set $\{0,1,2,3\}$ that has the greatest probability. Both the full and restricted models correctly classify the two firms with Intensity = 0. Both models underpredict the

TABLE 3
Ordered probit predictions of intensity.

Actual	Predicted				
	0	1	2	3	Total
0	**2**	0	0	0	2
1	0	**6**	9	2	17
2	0	7	**14**	7	28
3	0	0	9	**12**	21
Total	2	13	32	21	68

Notes:
Cells are the frequencies of actual and predicted outcomes. The predicted outcome has maximum probability. The results correspond to the model reported in the first column of Table 2.

TABLE 4
Predicted probabilities of intensity levels for variations in Tobin's q and leverage.

Intensity	0	1	2	3
30% below avg	0.0004	0.1072	0.4013	0.4910
20% below avg	0.0008	0.1498	0.4395	0.4099
10% below avg	0.0015	0.2021	0.4639	0.3325
Average q	0.0029	0.2636	0.4719	0.2616
10% above avg	0.0054	0.3326	0.4626	0.1994
20% above avg	0.0095	0.4063	0.4371	0.1471
30% above avg	0.0161	0.4809	0.3981	0.1048
30% above avg	0.0017	0.2105	0.4661	0.3218
20% above avg	0.0020	0.2275	0.4694	0.3011
10% above avg	0.0024	0.2452	0.4714	0.2810
Average leverage	0.0029	0.2636	0.4719	0.2616
10% above avg	0.0035	0.2826	0.4710	0.2429
20% above avg	0.0041	0.3022	0.4687	0.2249
30% above avg	0.0049	0.3223	0.4650	0.2078

prevalence of Intensity = 1 in the data, while overpredicting Intensity = 2. Presence of the strongest measures – noted by Intensity = 3 – are underpredicted, with 12 of the 21 firms being correctly classified.

To consider the implications of the model of Intensity further, in Table 4 we present the predicted probabilities that a firm will fall in Intensity categories $\{0,1,2,3\}$ as a function of variations in each of the explanatory variables. The bordered row in each

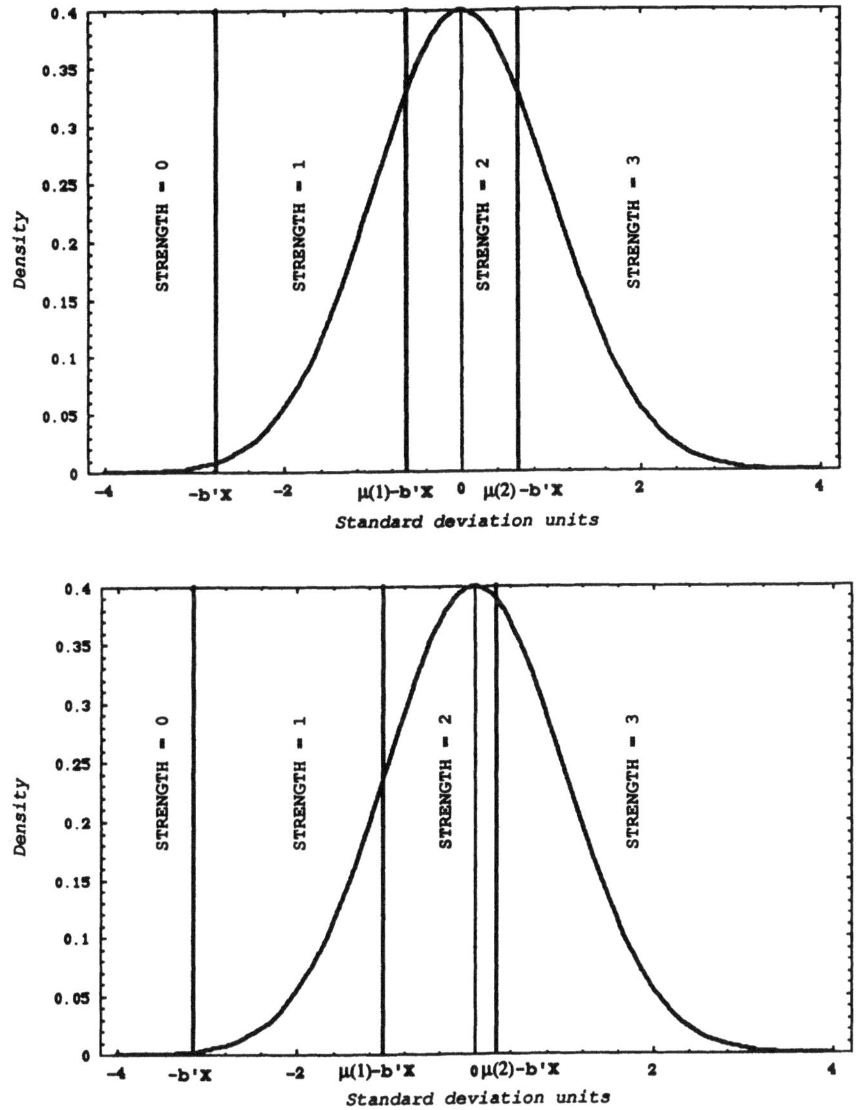

Fig. 1. Predicted probabilities of intensity levels. (a) Evaluated at point of means. (b) Evaluated at 20% below average q, mean leverage.

block is the predicted probability distribution at the multivariate point of means: in our sample, for average Tobin's q of 0.97 and average leverage of 0.43. The table shows how the probability distribution shifts with an decrease or increase in one of the explanatory variables, holding the other at its mean. For instance, the first row of the table considers the probabilities for a Tobin's q 30 per cent below the average value. The model predicts, in this case, a probability of 0.49 that such a firm will have Intensity=3: at least one of the severe anti-takeover defenses, accompanied by three or more of the less severe defenses, or at least two of the severe defenses. The distribution shifts markedly toward a stronger defensive posture with lower than average q, and vice versa. Figure 1 depicts how this shift takes place, illustrating how the thresholds between strength categories are displaced leftward when we move from average q levels to a q level 20 per cent below average. The results for variations in leverage are similar, if less marked. The probability that a firm will have Strength = 3 rises from 0.26 (with average leverage) to 0.32 for a firm with leverage 30 per cent higher than the average. These results predict a very strong response, in terms of the intensity of anti-takeover amendments, to either a low value of q or lower leverage. Since a low value for either of these financial characteristics is indicative of a greater probability of a raid (Servaes, 1991; Palepu, 1986), the presence of these amendments should not be expected to reflect shareholders' interests.

5. CONCLUSIONS

The presence of corporate takeover defenses – taken in combination – is shown to be clearly related to firm-specific factors such as Tobin's q and financial leverage. This result has been achieved by taking the qualitative information available on a firm's stance toward takeovers and applying a fairly loose classification to generate an ordinal index. The ordered probit technique is then used to relate the level of the index to a set of explanatory factors.

The multi-dimensional nature of many aspects of corporate behavior suggests that this technique might be useful in many aspects of empirical research. Since we are often unable to quantify the magnitude of a firm's stance toward the markets in which it participates, this method may have wide applicability. Indeed, although we have focused on corporate decision-making, the method would clearly be applicable to the analysis of countries, industries, or individuals.

REFERENCES

Brainard, W., John Shoven, and Laurence Weiss, 1980. "The Financial Valuation of the Return to Capital". *Brookings Papers on Economic Activity* **2**, 453–511.

DeAngelo, Harry, and Edward M. Rice, 1983. "Anti-takeover Charter Amendments and Stockholder Wealth". *Journal of Financial Economics* **11**, 329–359.

Greene, William, 1990. Econometric Analysis. New York: Macmillan Publishing Co.

Greene, William, 1992. LIMDEP Version 6.0 User's Manual. Bellport, NY: Econometric Software, Inc.

Jarrell, Gregg and Annette Poulsen, 1987. "Shark Repellents and Stock Prices: The Effects of Antitakeover Amendments Since 1980". *Journal of Financial Economics* **19**, 127–168.

Jensen, Michael C., 1986. "Agency Costs of Free Cash Flow, Corporate Finance, and Takeovers". *American Economic Review* **76**, 323–329.

Klock, M., C.F. Thies, and C.F. Baum, 1991. "Tobin's q and Measurement Error: Caveat Investigator". *Journal of Economics and Business* **43**, 241–252.

Lang, L., R. Stulz, and R.A. Walkling, 1989. "Managerial performance, Tobin's q and the gains from successful tender offers". *Journal of Financial Economics* **24**, 137–154.

Linn, Scott C., and John T. McConnell, 1983. "An Empirical Investigation of the Impact of Antitakeover Amendments on Common Stock Prices". *Journal of Financial Economics* **11**, 361–399.

Malatesta, Paul H., and Ralph A. Walkling, 1988. "Poison Pill Securities". *Journal of Financial Economics* **20**, 347–376.

Morck, Randall, Andrei Shleifer, and Robert W. Vishny, 1988. "Characteristics of Targets of Hostile and Friendly Takeovers", in *Corporate Takeovers: Causes and Consequences*, Alan Auerbach, ed. Chicago: University of Chicago Press.

Palepu, Krishna G., 1986. "Predicting Takeover Targets: A Methodological and Empirical Analysis". *Journal of Accounting and Economics* **8**, 3–37.

Pugh, W.N., Page, D.E., and J. S. Jahera, Jr., 1992. "Antitakeover Charter Amendments: Effects on Corporate Decisions". *Journal of Financial Research* **15:1**, 57–67.

Rosenbaum, Virginia K., 1986. "Takeover Defenses: Profiles of the Fortune 500". Washington: Investor Responsibility Research Center, Inc.

Ross, S.A., 1977. "The Determination of Financial Structure: The Incentive-Signalling Approach". *Bell Journal of Economics* **8**, 23–40.

Ruback, Richard, 1988. "An overview of takeover defenses", in *Mergers and Acquisitions*, Alan Auerbach, ed. Chicago:University of Chicago Press.

Servaes, Henri, 1991. "Tobin's Q and the Gains from Takeovers". *Journal of Finance* **66:1**, 409–419.

Thies, Clifford and Thomas Sturrock, 1987. "What Did Inflation Cost Accounting Tell Us?" *Journal of Accounting, Auditing and Finance* Fall 1987, pp. 375–391.

Zavoina, T., and W. McElvey, 1975. "A Statistical Model for the Analysis of Ordinal Level Dependent Variables". *Journal of Mathematical Sociology* **4**, 103–120.

List of Contributors

V. Eldon Ball
Economic Research Service, U.S. Department of Agriculture, Washington D.C., USA

Ravi Bansal
Department of Economics, Duke University, Durham, NC, USA

Christopher F. Baum
Department of Economics, Boston College, Chestnut Hill, MA, USA

C.R. Birchenhall
Department of Econometrics, University of Manchester, Manchester, England

Ismail Chabini
Centre de Recherche sur les Transports, University of Montreal, Montreal, Canada

Atreya Chakraborty
Lemberg Program in International Finance, Brandeis University, USA

Gregory C. Chow
Department of Economics, Princeton University, Princeton, NJ, USA

June Dong
Department of General Business and Finance, School of Management, University of Massachusetts, Amherst, MA, USA

Omar Drissi-Kaïtouni
Centre de Recherche sur les Transports, University of Montreal, Canada

Michael Florian
Centre de Recherche sur les Transports, University of Montreal, Montreal, Canada

A. Ronald Gallant
North Carolina State University, Department of Statistics, Raleigh, NC, USA

William L. Goffe
Department of Economics and Finance, Cameron School of Business, University of North Carolina, Wilmington, NC, USA

A.J. Hughes Hallett
Department of Economics, University of Strathclyde, Glasgow, United Kingdom

Charles Hallahan
Economic Research Service, U.S. Department of Agriculture, Washington D.C., USA

Robert Hussey
Department of Economics, Loyola University of Chicago, Chicago, IL, USA

David Kendrick
Department of Economics, University of Texas, Austin, TX, USA

Yue Ma
Department of Economics, University of Strathclyde, Glasgow, United Kingdom

Anna Nagurney
Department of General Business and Finance, School of Management, University of Massachusetts, Amherst, MA, USA

Alfred L. Norman
Department of Economics, University of Texas, Austin, TX, USA

Albert J. Reed
Economic Research Service, U.S. Department of Agriculture, Washington D.C., USA

Berç Rustem
Department of Computing, Imperial College, London

Agapi Somwaru
Economic Research Service, U.S. Department of Agriculture, Washington D.C., USA

George E. Tauchen
Department of Economics, Duke University, Durham, NC, USA

Utpal Vasavada
Economic Research Service, U.S. Department of Agriculture, Washington D.C., USA

Index

FSC
www.fsc.org
MIX
Papier aus verantwortungsvollen Quellen
Paper from responsible sources
FSC® C105338

If you have any concerns about our products,
you can contact us on
ProductSafety@springernature.com

In case Publisher is established outside the EU,
the EU authorized representative is:
Springer Nature Customer Service Center GmbH
Europaplatz 3, 69115 Heidelberg, Germany

Printed by Libri Plureos GmbH
in Hamburg, Germany